GO METAL DETECTING

Published by EP Publishing Ltd, Bradford Road,
East Ardsley, Wakefield

ISBN : 0 7158 0735 8
First published 1981

Printed in England by G Beard & Son Ltd, Brighton and bound by Seawhite Bindings Ltd, Hove

Val Singleton and Marguerite Fuke

GO METAL DETECTING

E P Publishing Limited

contents

PART ONE
Val Singleton

Treasure !

Have you read *King Solomon's Mines?* If you have you will know that the story contains all the essential ingredients of a magnificent treasure hunt. Allan Quartermain and his friends are guided by an old and sketchy map; they undergo incredible hardships before they find the site; and they have to battle against the malevolence of the hoard's protectress, Gagool. Their first reward when they enter the treasure chamber is a collection of ivory and gold. Then they come to a small nook at the back of the room:

We hurried up to the nook, shaped something like a small bow window. Against the wall of this recess were placed three stone chests, each about two feet square. Two were fitted with stone lids, the lid of the third rested against the side of the chest, which was open.

'See!' Curtis cried hoarsely, holding a lamp over the open chest. We looked, and for a moment could make nothing out, on account of a silvery sheen that dazzled us. When our eyes grew used to it, we saw that the chest was three parts full of uncut diamonds, most of them of considerable size. Stooping, I picked some up. Yes, there was no doubt about it, there was the unmistakable soapy feel about them.

I fairly gasped as I dropped them.

'We are the richest men in the whole world,' I said.

Although Rider Haggard's book is fiction it is no stranger than stories of real treasures such as the tombs of the Egyptian Pharaoh, Tutankhamun. Perhaps the greatest treasure of modern times, this almost went undiscovered when the archaeologists began to lose heart. However, the persistence of Howard Carter was finally rewarded in November 1922 when, after six years of searching, he put a candle through a hole in the door of the tomb he had excavated. He later wrote:

'At first I could see nothing, the hot air escaping from the chamber causing the candle flame to flicker, but presently, as my eyes grew accustomed to the light, details of the room within emerged slowly from the mist, strange animals, statues, gold—everywhere the glint of gold.'

Africa has furnished our first two tales of treasure but wherever you go in the world you will hear stories of riches awaiting discovery. Every country has its own legends to tell about hoards hidden away – by pirates or robbers maybe, or in a sunken galleon such as the *Duque di Florencia,* part of the Armada fleet thought to be lying deep in Tobermory Bay off the west coast of Scotland.

The glitter of treasure has attracted many adventurers over the centuries and has even taken some of them to their deaths. The Spanish search for the gold of El Dorado brought out both the best and the worst in the *conquistadors* who faced the hardships of exploring inhospitable lands occupied by the proud Incas. Men have gone crazy in their search for treasure, have killed their partners and have endured the awful conditions of the nineteenth-century 'gold rushes' in the United States.

But is that what treasure-hunting as a hobby is all about? Of course it's not, and yet the possibilities of 'big' finds are there, and every now and then an amateur treasure-hunter does turn up a hoard of plate or coins whose sale considerably improves his or her savings account. To take Britain as an example: so many different races have occupied the island since the first metal-workers were here, there have been so many invasions, wars and skirmishes that it is no wonder that there are many collections of gold and silver to be found. Look through any copy of *Treasure Hunting* magazine and you will read a true story of a new valuable discovery.

For instance, in the May 1979 edition there is the story of Tony Langwith, an ambulance driver who found a ninth-century gold plaque in a river bank. The staff of Norwich museum reported the find to the coroner who declared it to be 'treasure trove'. This meant that Tony was paid full value for the find – £14,400 – while the plaque itself was bought by the British Museum and is now on display there. Incidentally, I have earlier hinted at the greed of some treasure-hunters; it is pleasant to be able to record that Tony donated half his reward to the Brandon Playing Fields Association.

What are you Going to Find ?
Until now I have concentrated mainly on the classic treasures – gold, diamonds and sunken galleons. Well, gold and silver are certainly to be found in Britain and other countries, and you only have to think what life was like five hundred years ago to realise why. In those days there were no banks or cheques as we know them. If a man or woman had any money they would most likely keep it in the form of coins or plate which were made of gold or silver to their full face value, unlike today's coins which are a sort of token.

Tony Langwith's find – a ninth century gold plaque, worth over £14,000

Portuguese coins and a Spanish cob (bottom right), found in North Devon after a storm

There was then no police force or 'Securicor' to protect valuables, and as you will know from the stories of Robin Hood, an unaccompanied merchant or friar placed himself at the mercy of thieves once he set out on a journey. Towns, too, had robbers eager to seize whatever valuables a stranger might be carrying. Because of this, before a traveller entered a town he would often choose a hollow tree or suitable boulder and hide some of his valuables there. For many reasons the owner might afterwards be unable to reclaim the hoard; it would remain there waiting for a future finder to come upon it by accident or design.

On a more modest level, many are the single coins that have been accidentally dropped over the centuries. If you try in some way to calculate how many coins a person might lose in a lifetime, and then multiply that by the number of people there have been in a particular area, you reach a staggering figure for the total of old coins you should be able to find. Some of these are valuable, some are not—but they are nearly all interesting and, as my friend Jeff Short told me, many of them have a story to tell. Take, for instance, the 1746 silver crown I saw in Jeff's collection. A large coin, its most striking feature was the word 'Lima' under the monarch's head. The name of the capital of Peru was included on the coin's design to indicate that it had been struck from Peruvian silver once taken by the Spaniards. This silver was in turn captured by the British Admiral Anson and other privateers during their circumnavigation of the world in 1740-44 and brought back to Britain. And there it was, found by Jeff and worth about thirty pounds.

Please don't think that all coins and medals either cost or will sell for that amount. You can build up a fascinating collection quite quickly and cheaply, particularly if you are finding coins to sell, swap or keep. But while we are on the subject of the value of coins let me mention one other thing that Jeff told me. Among the many interesting objects in his window was a box of coins and ends of coins; the sort that you can buy from a dealer for a couple of pounds. Included with those was one mis-shapen, battered thing with a few marks and a ring of raised points. Jeff had a second look at it—yes, it was more than it seemed. An enquiry with an expert confirmed that it was a very old Celtic coin worth twenty times the value of all the other odds and ends put together.

This typified for me one of the first things to know about treasure hunting: there are many things that have value, whether they look attractive or not. If I were to take up this hobby the first thing I would do would be to try and familiarise myself with all the strange old objects that people collect. Books give a lot of advice on this and you can also go round antique shops,

A silver denarius of Vespasian,
found by Jeff Short

A George III 1816 shilling

A very rare 13th century seal, found in a motorway picnic area
in Kent. It belonged to Sister Ellen, who was in charge of a large
convent in the area

collectors' fairs or just keep your eyes open for newspaper articles describing auction sale prices. Did you know, for instance, that some old bottles will sell for hundreds of pounds each?

The great popularity of bottles is a relatively recent phenomenon. There are several books available from the library giving advice on how to find these, and essentially what you have to do is locate a Victorian rubbish dump or otherwise the uncleared cellars of certain types of shops – those that were once chemists or wine merchants. In the case of the dump you then have to dig down and sift through the soil to extract what once was regarded as rubbish but is now valuable. Personally I wouldn't enjoy doing this at all, but on the other hand I can see the charm of the bottle-collector's world, with its oddly shaped 'Codd's' and 'Hamiltons'.

Besides bottles you should be able to unearth stone jars, china ornaments, pot lids and clay pipes; once cleaned, many of these are surprisingly attractive. Perhaps today's aerosols and plastic beakers will be the next century's collectors' items, but somehow I doubt it. Go to an antiques fair and look at the bottles which will be on display there – you will see beautiful shades of blue and green, all shapes and sizes, and many bearing their manufacturer's name and the contents of the bottle. These are works of craftsmanship. Build up a collection of twenty or thirty and you will have a display that will give pleasure to all those who come and see it. And once again, your bottles and pots all have stories to tell which you will be able to work out.

I have said how easy it is to lose coins; equally, rings, brooches and small ornaments have been lost in their millions. A display of these can be most effective, ranging from military badges to jewellery. The places to look for these are where crowds have gathered – on beaches, for instance, or at a fairground site, or even at the site where a gallows once stood.

What else is 'treasure'? Well, I soon realised that the word covers a much wider range of objects than I had first imagined, that many of these are easy to come across, and some are not particularly old. For example, Marguerite Fuke, who writes the second part of this book, has built up a collection of horse brasses and ox shoes over the past couple of years, all of them found on her treasure-hunting expeditions. But quite definitely old is the arrowhead found in perfect condition by Jeff Short, and still in the hands of the museum to which he gave it for identification. He found this just by chance, as he walked across a newly-ploughed field.

Treasure usually means objects that have been made or collected by man and then deposited or lost. However, some forms are more natural, such as the semi-precious stones which can be found on beaches and in rock outcrops, and then polished up to be used in rings and brooches. This aspect of the hobby is closely related to fossil-hunting which really deserves a book of its own.

How do you go about finding the objects? I have mentioned two ways—by digging in old dumps and just by keeping your eyes open as you go around. But there is a third and, I think, more interesting way which has been developed over the last dozen years and to which most of this book is devoted.

My First Go with a Metal Detector

We have seen what a wide range of objects is meant by the word 'treasure' and also that the first thing a hopeful treasure hunter should do is to make him or herself aware of the range. The next step is to use your spare time most effectively to find the chosen objects.

There is a delightful Welsh story that relates how you should look for treasure. It was generally believed that hawthorn trees alone on a mound in the middle of fields marked the site of a warrior's grave. A long time ago, in the centre of a field near Llantrissant, Glamorgan, there was a very old hawthorn. A man living nearby had heard his great-grandfather telling a story of treasure buried in places like that, and decided to try his luck. He must first gather some springwort and forget-me-not flowers and leaves, and make them into a girdle, which had to be worn around the waist next to the skin. A sprig each of the herb and the flower were to be worn in the hat. Another condition to be fulfilled was that each time treasure was taken away the sprigs of the herb and flower worn in the hat were to be left as an acknowledgement of the transaction.

The man dug beside the hawthorn tree, and soon found a way down past the warrior's grave to a large cavern, where there were many human bones, beyond which he saw high heaps of gold. The entrance from the grave to this cavern was hidden by a stone carved with strange inscriptions. The man went for some months to the spot and, unknown to anybody, secured as much treasure as he could carry. A year passed, and the man grew very rich. However, one night he brought a load of gold that was heavier than usual, and in his eagerness to be quickly home with it he forgot to leave the herb and flower in acknowledgement of the treasure. When he next went for the treasure, he could not remember in which field the old hawthorn grew, and was never able to find it again.

'Right,' I said to my treasure-hunting guide, Jeff Short, 'is this really how you go about it?'

'No,' he replied, 'although an isolated tree might not be such a bad place

Treasure hunting the modern way – with a metal detector

to look, because it could indicate the site of an old building. Indeed, this autumn I intend paying special attention to a wooded mound in my local farmer's field.'

'A farmer's fields? Doesn't he mind you going over his land?'

'Oh, he's quite happy. I have his permission and have agreed that my partner and I will share good finds fifty-fifty with him. We are also covered by an insurance scheme which means that the farmer will be repaid for any accidental damage. I think he is only too pleased for us to remove anything that might harm his plough because this is all land that was used by the Army during the last war. A couple of weeks ago I located one of their trenches which was covered in corrugated iron and concrete—distinctly nasty if any deep ploughing is going on.'

I had better explain that Jeff is one of those people who have benefited from the great fillip given to treasure hunting by the development of metal detectors. The forerunners of these machines were the wartime mine detectors but their everyday popularity was assured once transistors were able to replace the heavier and more fragile components that were at first used. You can now hire detectors for a couple of pounds a day, or buy reasonable ones for under £30. The operating licence (supplied by the dealer) is quite cheap, and from then on the hobby costs nothing apart from replacing the odd broken trowel.

Jeff was a busy insurance executive three years ago when he bought a detector on impulse. He found a coin first time out and since then has added another three thousand to his initial find.

'I wasn't just hooked on the hobby,' he told me, 'my life was changed by it. I packed in the insurance job and started a shop in Ipswich for treasure-hunting enthusiasts.'

Jeff is also secretary of the Ipswich and District Metal Detectors' Club. One of the members, incidentally, is the Tony Langwith I mentioned earlier. He used a detector to find his ninth century gold plaque which is now on show in the British Museum.

After we had had our coffee Jeff packed some machines into the car and we set off so that I could find out what it was like actually using a detector. While we drove through the pleasant, flat fields of coastal Suffolk Jeff told me something about his big project, one based on historical research. It seems that a thousand years ago the coastline around Felixstowe was different from what it is today, so that flat-bottomed boats could navigate to ports which later became silted and which are indeed buried deep under those same fields and roads across which we were driving. The name for the major port of the region—Goseford—lived on until the eighteenth-century, though it was probably then two hundred years since the port had declined in favour of Woodbridge. Now it was almost impossible to believe that the small ditches and streams had once been the creeks leading to Walton Castle where Edward I had assembled a fleet.

'Goseford' was not a large town in today's sense; it was a collection of small townships including Falkenham, Candlet and Gulpher, and their locations are still something of a mystery to archaeologists and historians. Jeff was using his detector as a means of testing whether the historical evidence was true; somewhere under the fields he expected to find the metal objects which show where Gulpher had been situated. As the township was founded in the Roman era or earlier, prospects of an exciting find could be very good indeed, remembering that it was in this same river estuary that the Sutton Hoo burial ship was found.

So that was why we were now parked on the edge of a field of young sweetcorn, miles from anywhere except an isolated farm. As Jeff pointed out, the noise was deafening: pigeons, blackbirds, warblers, all singing undeterred by the occasional crack and echo of a rook scarer. I was put in mind of birdwatching and angling. For both of them comfortable old clothes is the order of the day; it was sunny so Jeff wore just jeans and a shirt, and around his waist a belt with two secure pockets. One of these was for good finds, the other for 'rubbish' which he removed from the site – part of his pact with the farmer. What else? Ah yes, around his right knee a skate-boarder's pad because, as I was about to discover, one's knee comes in for heavy use, so it needs some protection from dirt and broken glass.

Out of the back of the car Jeff produced a couple of metal detectors, one of which when new would cost about £30, the other £50 or £60. The cheaper one was the model Jeff had used to find all those coins of his. Its only disadvantage seemed to be that it was slightly more difficult to keep 'in tune' than the more expensive model which tuned itself at the press of a button. Details of how these machines work and what such technical things as 'discrimination' and 'ground effect' mean Marguerite Fuke explains later in this book. For me it was enough to know that I held my detector by the top of the handle and swept it slowly right and left across the ground in front of me. Normally a tuned detector emits a light buzzing sound. When it passes over metal the buzzing increases in intensity and you know that somewhere down there is a 'find'.

When we practised on the edge of the field my machine quickly gave off the unmistakable intense buzz. 'Aha,' I thought, 'gold so quickly. Marvellous!' Jeff's belt included a little hook from which hung a long-handled trowel. Using this he carefully scraped the soil from the area where

16

Farmland can yield many finds – above is an Anglo-Saxon buckle found in Norfolk

the detector sounded; yes, there it was just below the surface, my first discovery–a ring-pull from a beer can. Well, one is allowed one dud find; gold next time perhaps. We carefully refilled' the hole and trod it down so that the ground was in exactly the same condition as when we first started to dig. The ring-pull was placed in Jeff's rubbish pocket, and then we set off on our proper search across the field.

Bullets and Broken Horseshoes

Earlier on I mentioned birdwatching and angling, those two hobbies enjoyed by people who like the open air and possess infinite patience. The same qualities are needed by the treasure hunter; at least, so it seemed to me as we walked slowly forward between the rows of this vast field of sweetcorn. Keeping the head of the machine as low as possible so that it just brushed the earth we swept right and left one pace at a time. Actually, keeping the head low against the soil was quite difficult, because the plough furrows meant that the level was uneven, and also I did not like to bend the tops of the young plants. Whenever the machine came away from the soil the buzzing intensified slightly as a result of 'ground effect', but I soom came to tell this from the sharper note indicating metal.

The machines are not heavy and they are well balanced so that using them really is as simple as the manufacturers claim. Using them well is no doubt a matter of practice. The next time the distinctive note sounded I knelt down and used my own trowel to extract the find. This was more difficult than I expected. The soil was soft and dry, so there were no problems there, but after I dug a few lumps out I could still see no metal. I placed the machine over the hole; it buzzed again so that I knew that the metal was still be be dug out. A few more trowelfuls and next time the machine kept just to its continuous faint drone over the hole. Somewhere amongst the clods of soil around the hole was my object.

Guided by Jeff I picked up a lump of dug soil and held it over the head of the machine. Immediately it buzzed loudly. So then I sorted through the clod piece by piece, but still without finding metal. Surely the find could not be so small that it was invisible!

'Try putting your empty hand over the detector head,' said Jeff. I did, and the buzzing came again. Then I looked at my hand and saw on one finger the ring I wore as a matter of course and had completely forgotten. Object lesson number one: keep all other metal objects away from the detectors.

I now tried bringing clods of soil to the detector with my other, ringless, hand. Third time, it buzzed, and there in the centre of the soil was my second find: the metal cap to a cartridge case; value for collector or treasure hunter–nil. It was quite hard to spot the cartridge case in the soil; presumably this is why we find so few valuables when we dig in the garden–covered in dirt they pass unnoticed.

As we made our way across the field our detectors signalled frequently, and each time mine buzzed I felt a distinct thrill of anticipation. Quickly we unearthed quite a collection of–well, junk, I suppose, consisting of cartridge case butts, bullets, tractor bolts and broken horseshoes. These were all taken away for the dustbin. Retained to be cleaned and further examined were several military buttons and my one coin, a 1917 penny. It only took a couple of hours to gather this assortment so I am quite willing to believe that a longer search in a better spot could produce an exciting collection.

What are 'better spots'? Marguerite Fuke goes into this in some detail on pages 60-65, and there is no doubt that choosing the right site is the secret of successful metal detecting. One that I found and would like one day to go back and examine was in a small Suffolk village. To the south of the village was a flat marsh of reeds and poplar trees across which ran the main road to the nearest town. The village itself was on slightly higher ground and winding around it was the River Waveney. It was while I was gazing down into the river from the road bridge that I met an elderly village lady walking her dog. Soon she was telling me about the bridge:

'New this one is, built 1921. See those piles in the river, that's where the old one was, wasn't it? My uncle, he lost a gold watch building this bridge, his jacket fell in the water didn't it? And see along there to the right, that's where the old ford was! The road went along by those poplars and up against the church.'

I remembered this story and repeated it to Jeff Short who agreed that the old ford would be a good place to research. Metal detectors work perfectly well under water–and in the rain, incidentally–and there are various attachments and tools available to make river searches easier. Not that there aren't drawbacks; Jeff recalled one occasion when he was out with his family searching the track of a Roman road across some mudflats. He fell flat on his face while reaching for a clay pipe; his two children who were holding on to him also lost their balance and floundered around in the mud. Their dog, of course, rolled in it happily, thinking the whole thing was a game!

This sort of story told against yourself is a vital element of club activities, about which Jeff told me as we drove away from his medieval port. If you do take up this hobby I really do recommend that you join your local club. Besides hearing anecdotes, humorous or otherwise, you will hear talks by specialists such as coin experts or even bomb disposal officers. At Jeff's club there are various activities such as the 'Finder of the Month' competition, which Jeff finally managed to win with the Saxon arrowhead I mentioned earlier. They also go out together as a club, even abroad to search one of the French holiday beaches. An important venture is coming up shortly when the club are to take their detectors over the site of the new Ipswich southern by-pass, both before bulldozers move in and also when the top layer of soil has been shifted. Once the road has been laid the possibility of uncovering whatever historical remains there are will of course be gone for ever, so the Ipswich branch's work could be really important.

Above all, at the club you will meet enthusiasts who have had several years' experience (this hobby is so new that no-one has yet had very long experience) and who will be able to guide you on your first steps. Members are of all ages and from all walks of life. Thinking about 'walks of life', I asked Jeff what was the ideal life for someone who wanted to be a successful amateur treasure hunter. In his view you do best if you have a lot of weekday hours free; it is a good hobby for shiftworkers or for youngsters with school holidays.

And what about the future of the hobby? I had seen metal detectors that were too sensitive to be of real use to the beginner, and there are those that can discriminate between different types of metal so that you do not have to dig up every piece of iron you come across. Well, it seems that in deep-water engineering work the inspection of oil rigs in mud is being revolutionised by the use of sound cameras. These convert sound waves into pictures that look like X-rays, so you can actually 'see' the objects down there at the bottom of the river. Perhaps when technological advance cheapens these sonic cameras I shall get hold of one and take it back to my old ford in that Suffolk village.

Digging up the Past
Before you embark on this hobby I think it only fair to warn you that treasure hunters armed with metal detectors are not universally loved; indeed in some areas local bye-laws may prohibit their use. Why should this be? Surely metal detectors are quite harmless?

Not only are metal detectors in themselves harmless, they also have definitely beneficial uses, some of which were explained to me by Marguerite Fuke, who husband Peter works professionally to promote the wider use of these machines. I mentioned earlier that the treasure hunter's aid has been developed from wartime mine detectors, and you will probably know that advanced forms of these are used to search for weapons in trouble spots such as Belfast and Rhodesia. The police, too, make use of them in the rather macabre business of looking for bodies. On occasions amateur

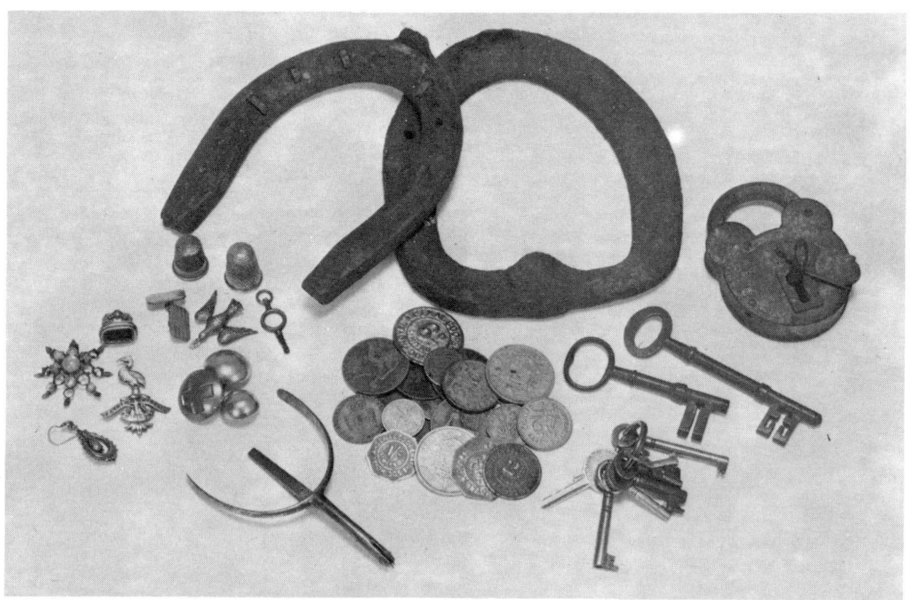

The sort of items you might find with a metal detector

Treasure hunting enthusiasts, young and old, enjoying a metal detector demonstration

Members of Ipswich and Colchester Detector Club assisting Essex police in the hunt for a murder weapon at Wrabness in October 1979

treasure hunters have been asked by their clubs to help with these searches or to seek lost valuables such as small items of jewellery.

Marguerite also told me about some of the less well-known uses to which the machines have been put; in Botswana they are used to ensure that no metal objects have found their way into the corned beef being processed for export. And did you know that detectors have found sheep buried in deep snow? The reason they can be used for this purpose is that most sheep are marked with a little metal ear-tag, and a good detector can pick this up.

So, to go back to my original question, why should anyone oppose the responsible use of these machines? An obvious source of opposition are property owners whose land you go upon without permission. Even birdwatchers, whom I would have thought to be the most harmless of enthusiasts, can incur the wrath of a farmer if they trespass on his fields; how much more likely is the anger of a farmer who suspects that you are digging holes in his crops. Indeed, even if the land looks barren to you the owner may still not wish you to go on it, and you must, by law, respect his or her wishes.

But apart from landowners there is a group of people with whom you may be in a sort of competition, since you both have an interest in digging up the past. To an outsider there sometimes appears to be a state of war between this group—the archaeologists—and the treasure-hunting fraternity. I have many reasons for feeling gratitude towards the profession of archaeology for the pleasure their exhibitions have given me on my journeys working around the world. That these people share the treasure hunter's excitements was revealed by Howard Carter's description of his discovery of Tutankhamun's tomb quoted on page 7. Here is another account, this time from nineteenth-century Britain. In 1885, on Queen Victoria's coronation day, a Mr Joseph Jackson crawled into a cave at Settle in Yorkshire. This is how he described his find:

The entrance was nearly filled up with rubbish, and overgrown with nettles. After removing these obstructions I was obliged to lie down at full length to get in. The first appearance that struck me on entering was the large quantity of clay and earth, which seemed as if washed in from without, and presented to the view round pieces like balls of different sizes. Of this clay there must be several hundred wagon loads, but abounding more in the first than in the branch caves. In some parts a stalagmitic crust had formed, mixed with bones, broken pots, etc. It was on this crust I found the principal part of the coins, the other articles being mostly imbedded in the clay. In the other caves very little has been found. When we get through the clay, which is very stiff and deep, we generally find the rock covered in bones, all broken and presenting the appearance of having been gnawed. The entrance into the inner cave has been walled up at the sides. In the insides were several large stones lying near the hole, any one of which would have completely blocked it up by merely turning the stone over. I pulled the wall down, and the aperture was now about a yard wide, and two feet high. On digging up the clay at about nine or ten inches deep, I found the original floor; it was hard and gravelly, and strewed with bones, broken pots and other objects. The roof of the cave was beautifully hung with stalactites in various fantastic forms and as white as snow.

In his deadpan account Mr Jackson omits to tell us that the coins were Roman, that there were delicate brooches, amber beads, bronze ornaments, finger-rings, armlets and bracelets. Jackson's cave, renamed 'The Victoria Cave', was thoroughly investigated by a committee of experts who identified the animal bones and classified the ornaments and coins. Gradually it emerged that the cave had been inhabited during the fifth century AD, in the so-called Dark Ages when the Roman armies had left the Britons to defend themselves against waves of barbarian invaders. A refuge in a time of lawlessness, it had remained undisturbed for 1500 years. The contents of this cave would have excited any treasure hunter, and the objects were valuable in themselves. More important for all of us, however, was the fact that those objects gave us information on how our ancestors lived – but remember that a number of experts were needed to extract that information and draw what we hope are the correct conclusions. So my own view is that when you get your machine not only should you keep off sites scheduled as being of archaeological importance (and you must do this by law), but you should also keep clear of the fields around the site unless you first come to an understanding with your local museum curators or archaeologists.

Now, just as each treasure hunter is an individual with his or her own views on how a treasure hunter should behave responsibly, so also are archaeologists individuals who may well disagree with each other on this matter. Some would like to see all metal detectors banned because of the abuses of a minority of enthusiasts whose actions would be frowned upon by the majority of those involved in the hobby. Others would wish to come to some sort of a working arrangement with treasure hunters so that each can benefit from the other's activity. On page 56 Marguerite Fuke gives a treasure hunter's view of how you should regard your responsibilities once you get that metal detector working. To get an opinion from the other side of the fence, I spoke to Tony Gregory of Norfolk Archaeological Unit, regarded as a 'moderate' willing to come to an understanding with treasure hunters; he was, incidentally, the curator to whom Tony Langwith took the gold plaque mentioned on page 8.

This is what Tony says:

'The points that we would like to see made on metal detecting are these:

1. All finds should be shown to archaeologists – the most important things are often the ones which look like rubbish, so save everything and get an archaeologist to examine it *all*.

2. Any gold or silver should be reported to the local coroner, because it should go to an Inquest to decide whether or not it is Treasure Trove – your archaeologist will help you to do this, but do it as soon as possible.

3. Always get permission from the landowner and the tenant (if applicable) before you go on to any land.

4. Find out if the land where you want to take your detector is part of an archaeological site or not; ask your local archaeologist and he will help you to find out. Obviously, steer clear of archaeological sites.

5. Record carefully where you find everything – in what part of which field – because when the archaeologist jumps up and down with excitement at the sight of the Roman brooch or whatever which you thought was a Victorian door handle, he will want to know exactly where you found it.'

Is This the Hobby for You ?
We have already seen what a wide range of objects is included in the word 'treasure', ranging from clay pipes to pieces of eight. To some extent anyone who keeps his or her eyes open for curious old–and sometimes not so old–objects is already a treasure hunter. Then the hobby could be divided into sections, some of which have been mentioned already. Locating and digging old dumps is the way to find bottles, jars and Victorian bric-à-brac; careful investigation of certain beaches and hillsides will yield semi-precious stones. A more adventurous branch of the hobby is the underwater diving for wrecks so well described by Jacques-Yves Cousteau in his book *Diving for Sunken Treasure.*

In many ways the last option seems to me to be the most exciting and the one in which I would love to participate. But this is an expensive pastime that is certainly not available to everyone.

So, if you can't dive for your treasure, why not take up the fourth option and try out a metal detector. The advantages of this hobby are that once you have your detector it is reasonably cheap. Like birdwatching you can pursue the hobby successfully everywhere you go, and in lots of different 'habitats' from the insides of old houses to remote beaches. Britain has been described as 'the real treasure island', and for good reason. There is just no end to the amount of valuables lost, hidden away or being washed up. But what goes for Britain also applies in other countries; you can pack your detector into a small carrier and use it in Portugal, for example, to uncover a whole different world of valuables.

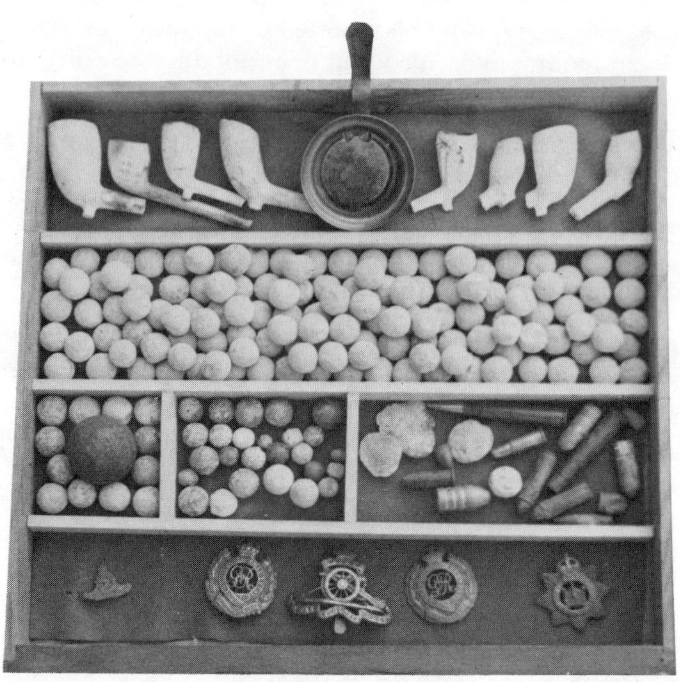

Two collections: clay pipes and militaria (above), and commemorative medals

Now, I ride horses for a hobby because I find it a complete relaxation from work, and that really is what I expect a hobby to give me. But I must admit that I can see that the added incentive of finding a small fortune and making money from a hobby must be attractive, so when you are choosing what to do you should remember that this is one pastime that might–and let me stress, it only *might*–bring you in a good bit of profit. One or two people have indeed become professionals who make their livelihoods from their detectors.

Another factor which I find attractive is that if you start using your metal detector now you are getting into the hobby fairly early; you will recall that this aspect of the hobby has only grown up over the past ten years. It should be relatively easy for you to become an expert and my guess is that there are some exciting technological developments to which you can look forward.

I have mentioned patience and persistence as being desirable qualities for anyone embarking on this pastime, but then isn't that true of any hobby if you really want to get the best out of it? Perhaps a better indication of whether you are likely to enjoy this hobby is how you feel when you turn up an old coin in the garden. Does it give you that thrill that Jeff Short still gets three thousand coins later? Do you rush indoors to clean off the grime and try to identify what it is?

Not only coins: Jeff's partner, for instance, specialises in rings, and you know that Marguerite Fuke prefers horse brasses. Younger hunters may relish collections of military buttons and insignia, or even the bullets that we were unearthing in their scores above the medieval port of Goseford. If any of these metal objects interest you then metal detecting could really be right up your street.

Various warnings are given throughout this book about how you must behave responsibly and not upset people unnecessarily or trespass on private land; and of course, how you must replace all your divots of soil. All this is really common sense of the sort that, again, applies to any hobby. One distraction you might not have to face in other pastimes is the horde of small boys which will spring up from nowhere and come and watch you fascinatedly as you search the edge of the village pond.

Is there much left to find? Surely someone will have got to the pond before me? 'Yes' is the answer to the first question and 'not necessarily' to the second. Jeff did rather ruefully say that he wished he had taken up metal detecting ten years earlier when he might have expected to find tens of thousands of coins per year; but don't worry, he still manages very nicely. It may well be the case that no one has yet looked at your pond, and even if they have they might not have been thorough. Added to that, detectors are

Solid silver thimbles from the late eighteenth and early nineteenth centuries

getting better (can detect smaller objects deeper) year by year and in any case it is extremely difficult to cover an area completely, no matter how clearly you work it out and how slowly you move. The reason can be understood from the following diagram:

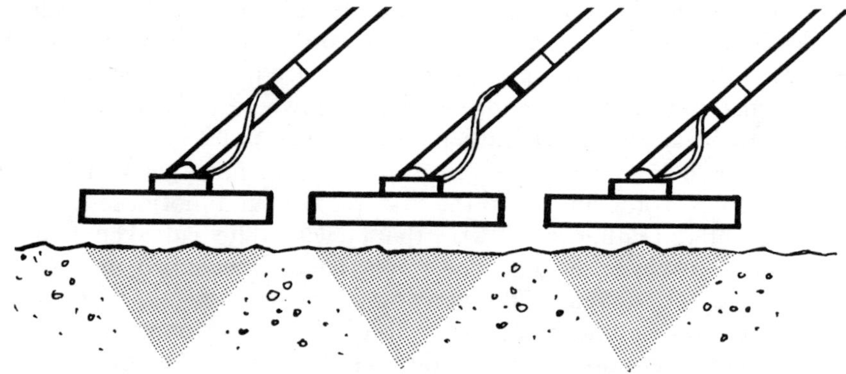

As the treasure hunter walks forward he tends to judge his step so that each sweep of the detector head just about overlaps the front edge of the previous sweep. But the effective search area narrows down to a point beneath the head, so this means that the unshaded area on the diagram has probably not been searched at all.

Well, as I discovered during my day with Jeff, detectors are easy to use and not too heavy, so that any of you reasonably fit youngsters should manage it with ease. There is now a full range available from several manufacturers so that everyone in the family should be able to find the right weight of detector and one at the right price. When you are thinking of buying your first machine don't neglect the second-hand market. Detectors are normally tough and long lasting, so the cast-offs of people who have become more engrossed in the hobby and have bought more advanced machines might well suit you for a start. In any case it is probably a mistake to spend too much money at first; make sure you are interested, learn thoroughly how to use your machine and only then go on to a more advanced model if you wish.

I know that to play football or to ride well needs not only physical abilities but also a good bit of intelligence and judgement. But I am sure that metal detecting is one of those pastimes that does your mind good in a way that football never could. Not only will you come to some sort of an understanding of people's habits when you begin to know instinctively which places (where on a beach, for example) are best to search, but you will also accumulate a good knowledge of the past. Just looking at a collection of found badges is an education in itself; how much more so is it to work out the stories behind coins as Jeff Short has done. You may come to know more odd historical facts than you ever learned at school.

So, if you have a natural curiosity, a collecting urge, reasonable patience and some free time during holidays, metal detecting should be for you. Go along to your local shop, hire a detector for a day and try it out. And don't forget, when you get hooked and buy your first machine be sure to join your local club or get in touch with other people already involved in the hobby. A whole world of local history is waiting for you to explore, there are thousands of odd items to collect and, who knows, maybe you will be lucky enough to find that crock of gold which eluded me in the beautiful fields of Suffolk.

Make sure the detector you choose suits your size!

PART TWO
Marguerite Fuke

Choosing your Metal Detector

So now you are convinced that metal detecting is the hobby for you and you are all set to seek your fortune. First, however, you must actually acquire the instrument of your fortune – an efficient metal detector. Initially the diversity of choice can seem thoroughly confusing, so before you go out to buy it is worth sitting down and asking yourself a few basic but essential questions.

First, how much are you able to afford? Will you have to save out of pocket money, wages from a part-time job, or is this to be a present from your parents? These factors may affect your price range but *not* the care with which you must make your selection. A standard model can cost anything up to £300 but it is certainly not necessary to spend that much on a first machine. I would not suggest that you go beyond £100 as an absolute upper limit and many machines will cost considerably less. Do remember, however, that generally speaking you can expect to find more with a more expensive machine and therefore recoup the cost more quickly – and lessen the risk of becoming bored. It also has a better resale value if you decide this is not the hobby for you. On the other hand, a cheap machine has to find less to pay for itself and your later finds will enable you to trade up to a more expensive machine when you are ready – many a successful treasure hunter has done just that. Some dealers may even be able to offer you a good second-hand bargain.

Other things you must think about are: how you intend to transport your machine – has it got to collapse down to go in a bag on the back of your bike? And are you a compact $7\frac{1}{2}$-stone or a 14-stone man-mountain? You may not like to say in a shop that the machine is 'too heavy for you' but if it is not comfortable to carry for any length of time you will soon stop using it.

So check the weight and how the machine balances in your hand. If you are still growing or are sharing the machine with a younger brother, girl friend or Dad, is the machine stem adjustable? Do you have any special requirements such as the need to plug in a hearing aid? Check the size of the jack socket and buy an adjuster if necessary. What type of land are you most likely to be searching? Will it be beaches, farmland or open moorland for instance?

After considering these questions, the next step is to collect all the literature put out by manufacturers. This will give you details of the latest developments and their advantages. But because sometimes it is easy to be confused by a salesman who may only want to sell you not the most suitable machine for you, but the one which gives him the biggest commission, I will set out here the basic terminology.

Metal detectors work by passing a rapidly changing current of electricity through a loop or coil of wire in the search head. This automatically sets up a magnetic field. Contact with the target object causes an inteference in this field which is 'detected' by the equipment.

Beat Frequency Oscillators (BFO) machines are the traditional mine detectors, though now considerably more powerful than the war-time versions. BFO models are the cheapest available and can even be purchased in kit form and built at home. They have certain drawbacks. Generally they are less sensitive and more affected by the response of the ground itself to their signal (this is known as the 'ground effect'), which makes them particularly unsuitable for beachcombing. They are dependent for their success on the operator's ability to distinguish the change in pitch of the continuous (and irritating) audible signal when a find is made, and upon his judgement on the likely composition of the target object. However, if this is all you can afford, do not despair. I have seen BFO users compete very favourably with more sophisticated equipment users in competitions – although I must stress that the users had had a great deal of practice. Another factor in their favour is that you may get a great deal of pleasure and satisfaction out of actually building your machine and using something you have built for yourself. So don't dismiss these models without some thought, for they could be the right start for you.

Induction Balance detectors (IB) utilise two coils. These are, effectively, a 'send' coil and a 'receive' coil. A find interferes with the balanced communication between them and is registered audibly. IB machines are susceptible to ground effect but this can be tuned out with some small reduction in sensitivity.

Metal detectors from oposite ends of the range. Above, a very basic induction balance machine. Below, one of the most sophisticated machines you can buy – an auto ground exclusion discriminator with auto tuning and mode selection

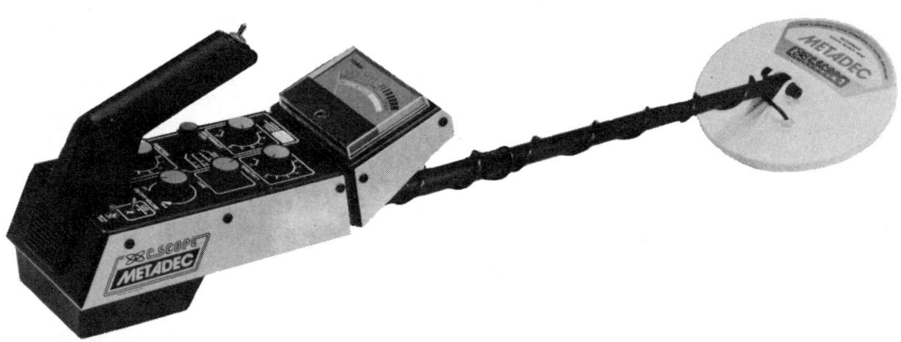

Transmit-Receive detectors (TR) is a term which is frequently used for IB detectors, but some manufacturers reserve the expression for those utilising a particular arrangement of search coils which is also called the 'wide scan' head.

Many of the slightly more expensive machines have a special discrimination facility which is designed to save you the chore of digging up objects such as silver paper all day. As with any 'tuning out' facility, this is only achieved by the loss of some sensitivity. If you choose a machine with this facility it is sensible therefore to search at the 'normal' level and only switch to discrimination when a find has been located, to check whether it is likely to be worth digging up. The major fly in the ointment is the ever-present ring-pull, from cans of all sorts of drinks. Tuning these out may well result in the tuning out of some kinds of gold rings, not to mention small bronze Roman coins. So it is up to you to decide, in view of the nature of the sites you expect to search most, whether this facility is what you want.

Very Low Frequency machines (VLF) are designed to counteract the effect of ground conditions and make discrimination easier. They are extremely sensitive to iron. There is a whole range of machines available, some with and some without discrimination. Their performance differs widely and they are being updated at regular intervals, so it is really essential for you to try out those within your price range when you are ready to buy.

Pulse Induction machines are without doubt the most sensitive and deep-seeking of machines. However, their very sensitivity renders them unsuitable for general treasure hunting, particularly for the beginner. As usual there are two sides to the question: if your interest is primarily archaeological then you may not mind digging up nails 8 inches deep, so this machine could be the one for you. They are certainly highly efficient at tuning out ground effect. I use one myself, but only for beachcombing or when working in conjunction with someone who is using another type of machine.

The nearest the metal detector industry can supply to a *Which* report is the annual *Treasure Hunting Manual* brought out by NPC Publications, Brentwood. Details of their machines are supplied by all the major manufacturers, which is useful for viewing the range of machines, but you must remember that these are not independent tests by an impartial organisation. *Treasure Hunting Magazine* does carry out its own tests and these are very helpful (provided the machines have not been amended since testing in the light of the results!). If you have already got to know your local Club you will have found that the members are a mine of information

– though, human nature being what it is, you will need to make due allowance for the fact that every member is absolutely convinced that his is the very best machine on the market.

After *what* should you buy, your next question is undoubtedly 'where shall I buy?' Detectors can be purchased from a number of different outlets. The major catalogues and some department stores are useful sources, because they usually have easy payment terms available; but their selection is very limited and so is their technical expertise.

The best plan is to go to a local dealer. Not only is he likely to be an enthusiast himself, but he will have, or should have, a wider range of makes available. These should include C-Scope, Savo White's, Fieldmaster, Garratt's, Fisher's and many more. The shop should have adequate facilities for comparing the performance of different machines. Check how easy it is to pinpoint a find: the smaller and neater the incision you have to make, the less effort you will expend over the course of a long day.

Get the dealer to test the machines you are most likely to purchase *under exactly the same conditions,* that is, all in air or all in a sandpit, with and without their discrimination facility, if any. (It hardly seems necessary to say that it is extremely unwise to buy a machine which the dealer is unable to demonstrate because, for instance, he has no batteries or they have gone flat through standing in the sunlight in the window, or because he simply has not taken the trouble to find out how to operate that particular model.) Take notice of the colour of the search head, for this is important since heat and light waves not only reduce the effective life of batteries but also affect delicate circuitry. These waves are attracted by dark colours and repelled by light ones.

Some expensive models are difficult to demonstrate in a shop since the surrounding metal may affect them because of their great sensitivity. The dealer will undoubtedly have access to some kind of yard or even a car park where you can test the performance of these machines. You will also need to check that the machine is sturdy and not likely to break if you accidentally lean on it, and that the cord around the stem is long enough to enable you to adjust the length of the stem without pulling the cord away from its connections.

Now is the time to make certain you can operate the machine you have decided upon. The normal method of searching is to adjust the stem length so that your arm can hang freely and easily with the search head parallel to and about an inch from the ground. Instructions for tuning each model, even within the same range, differ widely, but the general idea is to tune for maximum sensitivity on the 'normal' mode and then reduce the volume

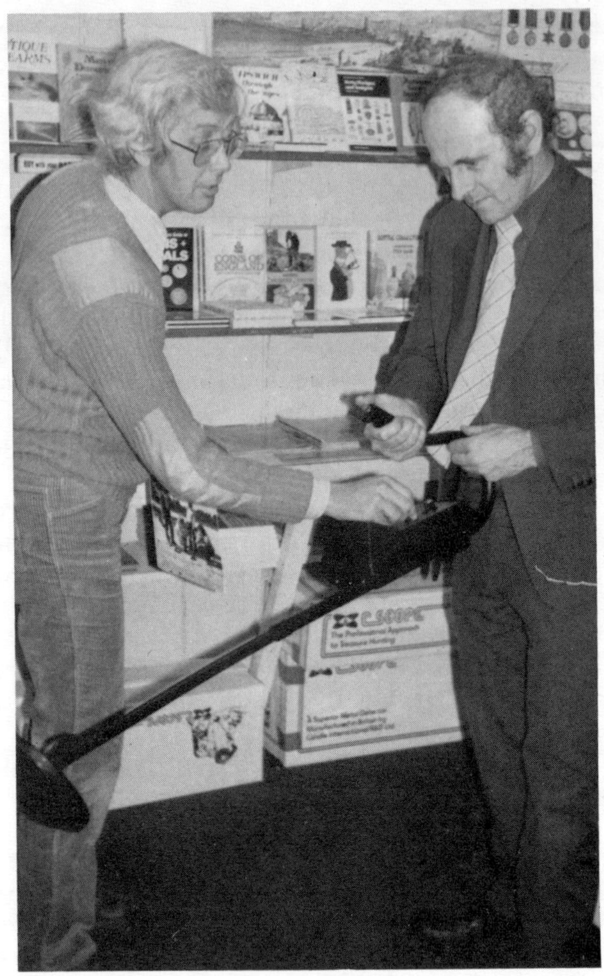

Demonstrating a detector to a customer

until it is so low you can almost not hear it. Any alteration in the signal will then attract your attention very easily. Make sure you can do all this before taking the machine out of the shop.

Just one final word on buying your machine – take your time. Don't let the dealer rush you. Buy when you are ready and not a moment sooner.

So at last you have got your machine home. As soon as you can, read the instructions carefully and start practising. It is a good idea at this stage to protect the search head with a clear plastic bag to keep it clean and undamaged. It should not take you very long to master the art of the steady, regular search swing and the location of your test objects – coins are probably the most convenient to start with. See if you can estimate the size – or depth if you have the opportunity of burying test objects – of your find by the difference in signal. When you are happy with your performance you can start to experiment with the discrimination mode; start by cutting out ordinary bottle tops to get the procedure right, then try silver paper (actually aluminium foil), iron nails, ring-pulls, etc. If you have chosen a non-discriminator model, listen to the signals given by different metals and see if you can tell one from another.

If you are not satisfied with your achievements go back to the dealer *as soon as possible* and insist that he either shows you where you are going wrong or proves to your satisfaction that the machine is not faulty.

Some Useful Accessories

There is only one golden rule when deciding what to take with you on a day's metal detecting: if in doubt, leave it out. You may well have to carry everything with you all day and, believe me, by the end of the day your 'essential' equipment will feel three times as heavy as when you set out.

First things first – what are you going to wear? Leave your pointed shoes at home and get yourself a pair of comfortable and really waterproof boots instead. Army surplus gear is ideal. A parka with a hood, and enough large pockets to contain everything else you need, is extremely useful – and it doesn't matter if you get covered in mud.

My detector has a carrying bag which has several useful pockets for my maps, tide table, compass, lunch, etc. My trowel fits into a special pocket – it is a slim-line trowel since there is no point in digging an unnecessarily large hole. In practice, I find a sheath knife is perfectly adequate for digging up a single coin. I am also experimenting with a special turf cutter but this is heavier and I don't take it if I am going to have to carry it around all day. My compass has been a constant companion since my Duke of Edinburgh Gold Medal Award days when I first got lost in Epping Forest! It is a Silva compass, lightweight, apparently indestructible and it makes map reading dead simple.

Another pocket contains a number of small plastic bags for protecting and isolating finds, and a book of raffle tickets. Yes, a book of raffle tickets. I find them the easiest way of cross-referencing finds with my usually muddy notebook in which I detail the exact spot in which I made the find. A large plastic carrier bag starts the day in the same pocket. This is to hold all the real junk I dig up – many a farmer has been finally influenced to allow me to search his land by my promise that I will faithfully collect all tin cans, etc., which could cause injury to his livestock or foul up his combine harvester blades.

For more civilised sites such as parks and lawns I also carry a small camera (as a precaution against false accusations of damage) and a plastic sheet to hold the excavated earth tidily and cleanly. This also makes the job of returning the earth much quicker and keeps my knees clean and dry.

When searching a large site such as an unmarked field I take a number of bright red pegs and plenty of string along to mark out my search lanes since nobody, but nobody, can walk in absolutely straight and parallel lines without guidance.

Specialised sites such as rivers and ponds call for specialised aids. They include long-handled scoops and sieves, both of which you can design and

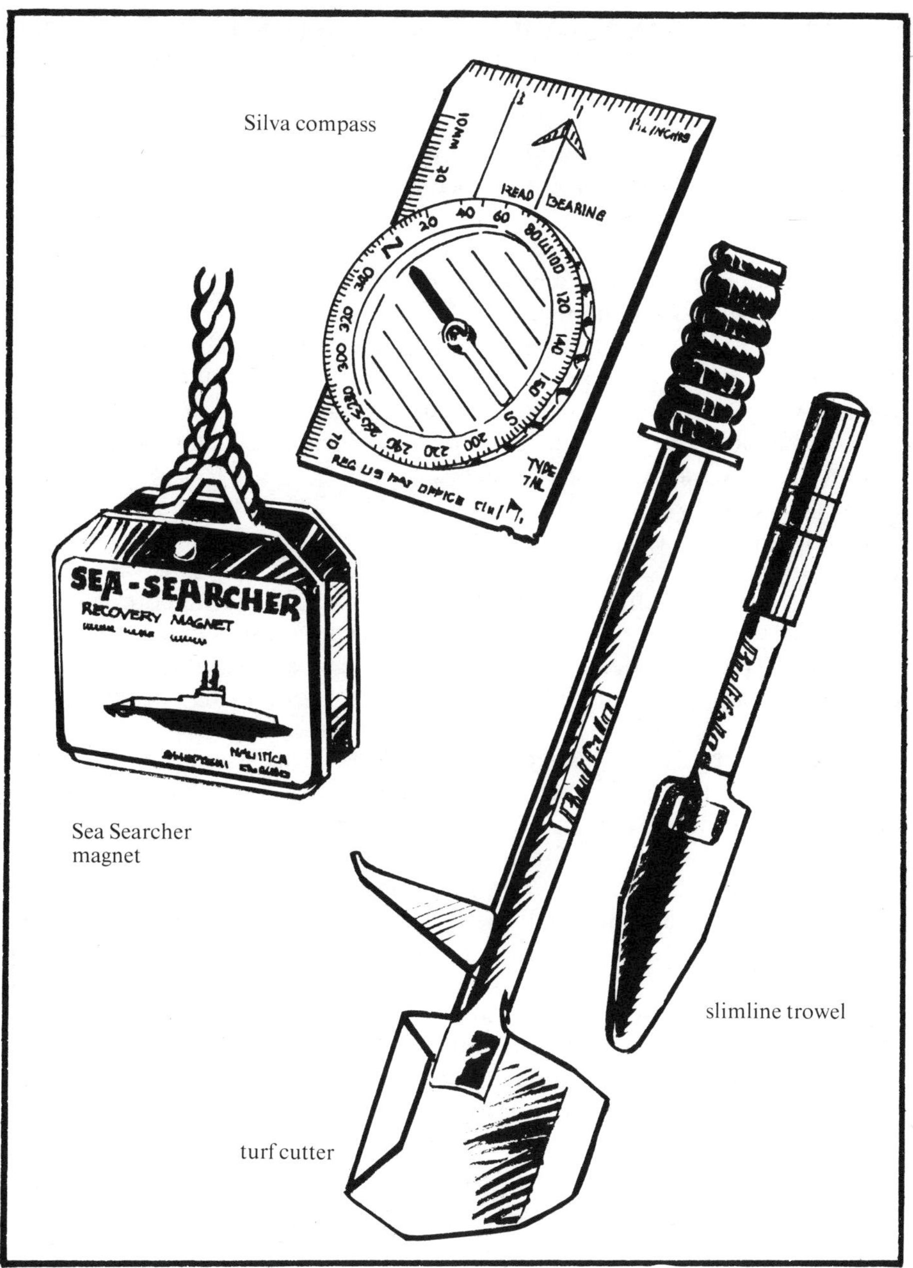

Sea Searcher
magnet

slimline trowel

turf cutter

make for yourself. The most popular components for a sieve are an ordinary plastic garden sieve with the inner tube from a car inflated and tied around the outside.

The only other essential aids are a completed Landowner Agreement Form – which I will talk about later – two sticking plasters for feet or fingers and a bar of your least favourite chocolate (it must be your least favourite otherwise it won't survive the day) for when home and supper seem just too far away.

It is a good idea to carry a whistle if you are contemplating an area such as Dartmoor where a sudden mist could leave you isolated.

Six calls on the whistle is a recognised distress signal and, if magnified by being blown across the top of your wellington boot, could bring half the British Army on exercise to your rescue! A slightly heavier duty than usual knife will enable you to construct a shelter if you are really lost, and a packet of waterproofed matches adds a touch of luxury.

Back at home I keep a number of other aids. The most important of these is undoubtedly my coin cleaner. This needs to be used with great care, for over-zealous cleaning can result in reducing the market value of a coin or even in cleaning off all remaining means of identifying the coin itself and leaving only a blank disc. I use an electrolytic coin cleaner obtained through the British Detector Society; other electrolytic cleaners, sonic cleaners and chemical cleaning kits are available but I have found this one to be most satisfactory.

My 'better' finds are displayed in a display case from Jim Chapman Detectors, Sandy, Beds., but there is no reason why any reasonably competent handyman (which I am not) should not make his own.

A 'Sea Searcher' magnet for all water work, detector legs for supporting my carefully tuned detector while I am digging and a battery check meter complete my own equipment list. Many other items are available for your Christmas and birthday present lists: additional search heads, covers for search heads, detector test pegs, larger digging tools, etc. Books alone can keep your relatives buying for years. The most important, even essential, books to have available are the first-class coin guides from Seaby of London. These are hardback books and hence not cheap, but they are well worth the investment. If you really can't afford them look for them in your library and, if necessary, ask staff to obtain copies for you.

How to Search a Site

Now you have bought your machine and decided what accessories are absolutely necessary to your efficiency and comfort and you are itching to get started. But just hold on a minute longer, for there are some points you need to be aware of before you set off.

Unless you are going to a completely isolated site miles from anywhere you have one special factor to consider. You and your detector are going to be the focus of all eyes. Some won't be very friendly, so polish up your smile before you set out, and make sure you have your Landowner Agreement Form, Code of Conduct, etc., conveniently to hand. Some of those interested will be men who used metal detectors during the war, who may want to take the chance of reminiscing about their experiences for hours – don't antagonise them, but try to explain tactfully that the cheaper BFO models now available are more powerful than the old back packs they remember and turn the conversation on to just what *you* are doing – and keep searching.

The other effective way to discourage most adults, if you are getting desperate, is to wear headphones and explain with smiles and sign language that you simply cannot hear a word they say. Unfortunately, this method will not work with the biggest crowd of onlookers – small boys. Small boys and girls will appear as if out of rabbit holes as soon as you unpack your machine. They are quite prepared to be either awed, respectful and enthusiastic diggers, or loudly derisive; so make sure you have really practised tuning your machine and finding planted objects in your back garden till you can almost do it blindfolded. On location with an audience is no time to sit down and work out how to put your machine together or to read the instruction book. Now is the time for a slick, self-confident display of the expert at work.

Next, take time really to look at the site and to plan your method of approach. Now is the time to identify areas for special attention, such as ancient trees which could be boundary markers: a group of three fir trees, for instance, could mean that this area was once set aside for drovers to stay in overnight on their way to market with their cattle or sheep. Look also for sheltered spots. The place you choose to eat your lunch out of the wind and sun could well be the very place where farm workers have munched their bread and cheese for two hundred years.

You must also check up on any possible dangers. Check the date on your tide tables, locate any quarries, wells, etc., marked on your map and make sure the farmer has not forgotten that you are coming and turned out his

Watch out for bulls!

prize Devon Red bull.

There is one very commonplace hazard which is often under-rated by those who have not suffered from it: cows. Cows and bullocks are extremely curious creatures – and playful in their own way. One 6-foot, 16-stone man who prefers to remain anonymous was once standing looking down into a hole in the ground watching the operation of a hydraulic ram attached to an adjacent trout pond. Only instinct made him turn round just in time to avoid being nudged ever-so-gently into the hole by a playful Charolais bullock named Napoleon. Another man found himself edged into a large muddy pond by a herd of curious cows and there he stood, up to his waist in sludge, holding his detector over his head, for some time before his farmer friend rescued him. This same man soon afterwards became the first Chairman of the British Detector Society. Generally speaking, however, bullocks are non-aggressive, but cows with their young may feel threatened for no apparent reason, so keep an eye on them all the time.

It is very important to be systematic in your searching; don't wander aimlessly around or you will miss some areas and search others several times. Remember no human being can walk in a straight line! Prove this for yourself by lining yourself up with a landmark on the other side of a large field, put on your headphones, switch on your detector and head for the landmark whilst searching. You will be surprised. Childhood adventure stories of explorers walking round in circles in deserts or jungles are absolutely plausible.

A newly-ploughed field is a god-send, with beautiful straight lines marked out for your convenience; or a field may be planted with, say, sugar beet, again in lovely straight lines. If you have a site like this make the most of it.

Not all sites lend themselves to subdivision, but where this is possible the most practical method is to use pegs and lines.

The most usual method of scanning the ground is to hold the detector with its head parallel to the ground and 1 inch above it. The detector is then moved as far to the right and left as the user finds comfortable. Try this at home and measure the distance between far left and far right. Then subtract several inches to allow for an overlap. This is the width you will need for your search lanes. I find red is the easiest colour to spot, or one of the more garish fluorescent paints – these are especially useful at dusk. Only two lines are needed as the first can be moved on to make the third and so on. Search lanes of this type were recently used by the Scottish police when they successfully located a stainless steel dinner knife which had been used to stab a man in a pub brawl and then buried point down in the ground.

When you pause for a breather, or at the end of the day, you will need to take careful note of the exact spot where you stopped, either by using a marker such as a handkerchief tied to a hedgerow or a particular formation of stones or by lining up your position between two or more outstanding features, say a church spire or a live elm tree. Even a tied-up tuft of grass, or a longish stick pushed into the ground, will do, to ensure that you don't miss any ground. I am sure you do not need to be reminded to remove all markers when finished with and not to use anything which could be dangerous to livestock. Even string constructed of man-made fibre can be fatal to a bird.

Different sites require different techniques. In woodland it may be difficult or dangerous to use pegs and lines but helpful to use chalk marks on tree trunks. If you are searching a ford you will need to cover some 20 yards on either side since fords move over the years and coins, or the contents of the inevitable overturned cart, may be washed downstream by the current.

In houses certain locations were certainly particularly popular for hiding savings. Under the stairs, the floorboards and alongside chimney breasts were favourite places. In a garden it does not pay to neglect any area as being unlikely. The great diarist, Samuel Pepys, records having given his wife a tremendous scold when she and her father admitted to having buried their treasure in the middle of their garden where any of their neighbours might have observed them! They lived to retrieve their hoard but thousands of others did not.

Wherever you are searching, the key is to ask yourself where you might have hidden something or lost something and what is the most practical method of searching for it.

Random searching
Some areas are covered several times, others not at all

Line and peg searching
(Only first two lanes marked.) Gives far better coverage of the area

approx. 2ft

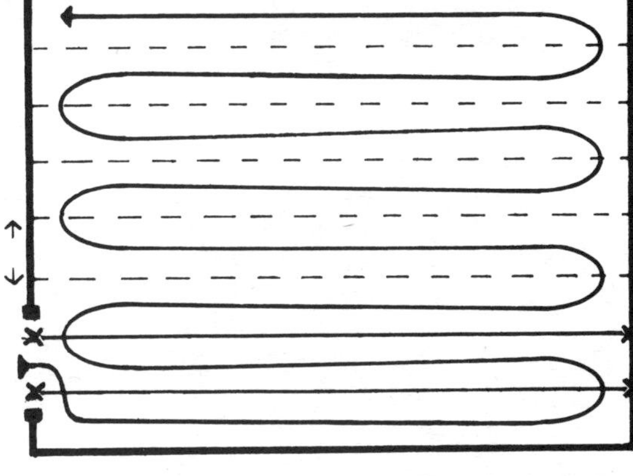

Lines and pegs can then be used along the area instead of across it, to give even more thorough coverage

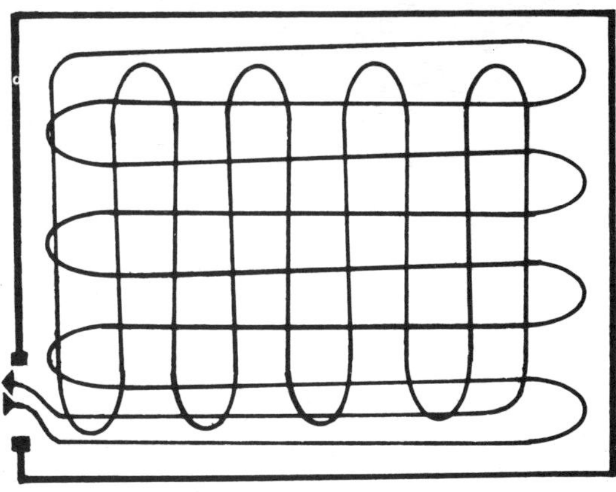

Dave Kitchen's four golden guineas

Techniques of Digging

First of all it is worth while considering the object of digging at all. To my mind it is to retrieve the located object with the minimum expenditure of effort and time, causing no damage to the object and leaving the site in an irreproachable condition. If you have a mental image of yourself with a spade over your shoulder like one of the seven dwarfs, forget it! The vast majority of finds are single coins within 6 inches of the surface.

One fortunate treasure hunter, Dave Kitchen, recently found four golden guineas buried on the site of some old farm labourers' cottages. Each was located separately and each was within about 4 inches of the surface. He dug out each very carefully and their excellent condition caused them to be valued at no less than £7,500!

Since you have selected your metal detector with great care you will be able to pinpoint a find extremely accurately before you start excavating. Practice will enable you to form a general idea of the depth and size of the object from the signal. A small sheath knife (or even a knitting needle or a screwdriver) can be used as a probe to locate the object physically and you may need nothing more than the screwdriver to extract the coin. The easiest way of removing the object is to dig out a plug of earth shaped like an ice-cream cone.

We all get very impatient to see what we have found but remember that the condition of an object may have a considerable effect on its market value. Had Dave Kitchen whacked one of those golden guineas with a spade he might – no, he certainly would – have knocked thousands off their potential value.

A large hole takes longer to 'backfill' – that is, fill up after you have taken the object out – than a small so the golden rule is 'It's smart to stay small'.

For slightly larger finds you will need to take up a turf, that is to say, you will need to remove a patch of grass with the roots intact in the form of a 'slice'. To do this just cut through the grass layer on three sides of the area and, using a knife or trowel, cut under the grass roots so that you can fold back a sort of trapdoor. Leaving one side still attached helps the grass to grow back quickly and minimises the amount the turf will shrink. Having cut your turf, run your detector over the area again to make sure the object is not in the turf itself, then you are ready to begin digging. I use a plastic sheet to hold the excavated earth wherever I am digging as it also helps protect my knees from damp or brambles or anything else I might kneel in. The sheet also makes it easier to return all the earth to the hole with the minimum amount of effort afterwards (I don't believe in wasting energy).

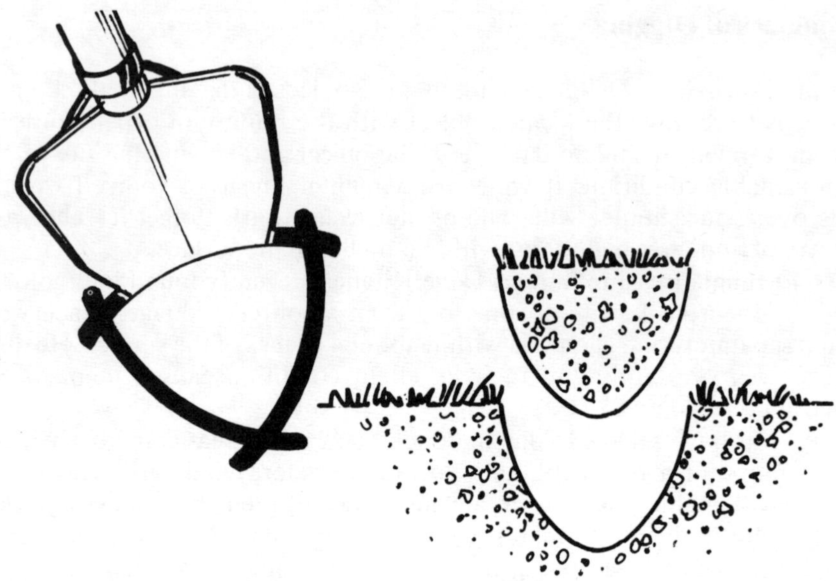

Removing a small plug of earth – first
make three trowel cuts

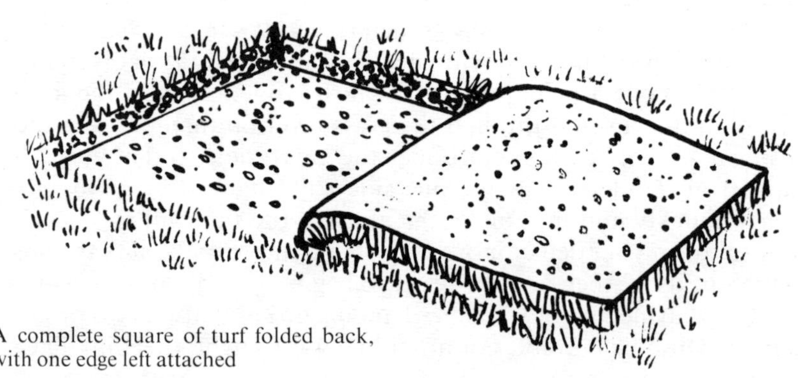

A complete square of turf folded back,
with one edge left attached

It is sound commonsense to scan the site at intervals as you dig, since it only too easy to remove an earth-covered object without noticing it.

When you reach your find, handle it very carefully. It might be very old and fragile. If it is obviously junk, such as a discarded beer can, throw it into your junk bag to be disposed of in the nearest litter bin later. Incidentally, many charities now collect the ring-pulls from cans so it is worthwhile saving them, since you will undoubtedly pick them up by the thousand.

If you think your find may be worth keeping, or if you simply cannot identify it, put it in your finds bag to study later at your leisure. I prefer to use a number of separate small bags to prevent items rubbing against one another and perhaps damaging each other. If, like me, you decide to keep all the horse and ox shoes – and there is a tremendous range of them – you will probably prefer to leave a box somewhere convenient and take the shoes to it since they are awkward to carry around. If you are going away from the spot remember to leave something to indicate exactly where it is so you can take up from the same place. Never leave your detector if there are livestock in the field.

I have placed great emphasis on the need to leave a search area in the condition in which you – or more to the point the owner – would wish to find it. If you wish to be invited back rather than chased off by a shotgun-wielding farmer you must treat land with the same consideration you would show towards the owner's house. Metal detecting as a hobby is still in its infancy and if we, the pioneers, are to continue to enjoy our hobby for the rest of our lives we must make a particular effort to be seen to be behaving responsibly, otherwise we will see the 'no metal detectors allowed' sign going up and we will have to miss out on a lot of enjoyment.

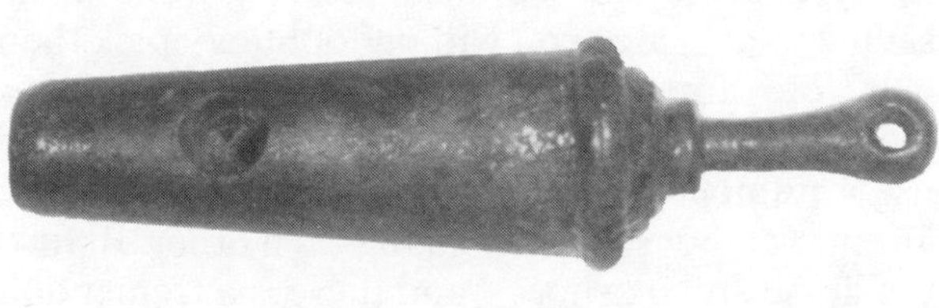

Some interesting finds: a Georgian 2 oz weight (above left); a 1750 hawkwhistle (above right); and a seventeenth century pewter jug

Victorian children's metal toys

How to Plot and Record your Find

In the middle of the last century the famous American outlaws Jesse and Frank James buried their ill-gotten gains in the middle of a desert while making a quick getaway. Unfortunately, their getaway was a bit too quick – they neglected to plot the spot as carefully as they ought. As a result the not-so-small fortune which was supposed to enable them to live in luxury for the rest of their lives still lies buried where they left it. Jesse was killed and his brother spent the rest of his life wandering round the desert looking for his lost fortune.

So be warned! Every month hoards, large or small, are being discovered in this country, proof of the fact that many people in the past have been equally careless. That find you almost junked many weeks ago could turn out, when you get around to cleaning it, to be valuable enough or interesting enough to make you go back and search the same area again more thoroughly, and you will kick yourself if you cannot remember where you found it. So make a careful note in your finds book of where you discover anything and everything.

You may not be the only one interested in the exact location of a particular item. Not necessarily valuable in cash terms, it may be of considerable interest to your local archaeologist who will need to know precisely – to the exact millimetre – where you found it. He will also want to know exactly how deep the object was, so make a note of that even if you are working in a ploughed field.

The relative positions of a number of finds may also be of tremendous significance. A collection of leaded lights or house nails, for instance, may indicate the site of an old building, and even things such as old tiles which you may come across could lead to the discovery of a hitherto unknown Roman villa.

Plotting finds can be a highly sophisticated business. However, unless you, or someone you know very well, just happen to be a surveyor I suggest you leave the theodolite at home (some clubs do actually use them on club outings) and concentrate on simpler methods.

The ideal situation for plotting finds would be a small field, exactly rectangular, running exactly north/south and empty of livestock. Regrettably, I know from experience that they do not exist in this country. So you must be prepared to adapt the basic methods to suit the circumstances.

The simplest method is that used in a field where you are using lines and pegs. All you need to do is measure the distance from the beginning of the lane, either with a tape measure or by counting the number of your paces from the beginning – you have probably watched show jumpers measure the distance between jumps by this method. (If you measure the length of your stride, which will vary according to whether you are walking on heavy, wet ploughed or light dry soil, you will find it is pretty consistent from day to day. Or you could usefully spend a rainy evening marking off your line in yards.) Next note down the number of the lane, that is, your first, second, etc., and the number of inches in from the line.

An infinitely more reliable method is to use compass bearings. This may take a little more practice but is much more efficient. Stand at your find and look about you for some easily recognisable landmarks. Those which can be found on your Ordnance Survey map are the best: gates, pylons, the corner of the field, for instance. Now measure either the number of degrees between each of three or more markers or the variation of each from north. (Remember that magnetic north and the grid north shown on the map will vary by a few degrees.) Make a note of the bearings and put a mark and reference number on a sketch map of the field, which can easily be copied from the Ordnance Survey map on the evening before. When you sort out your finds at home this point can be plotted more accurately on a map for future reference. This method is also useful when you stop searching for the day or the weekend. It can also be adapted for use on other sites such as beaches and rivers.

I regret to say that I have yet to find a really reliable method for the middle of, say, Dartmoor. There is a distinct shortage of landmarks and although I have used bearings on the sun, noting the time of day, and cross-referencing as best I can, my experiment was not very successful – if Christopher Columbus had relied on me to guide him across the Atlantic he would be still sailing round in circles! If you work out an infallible method please let me know about it.

Extract from 'Finds' log

Location : The Slips — Farm Ongar.
Date : June 1980.
Ground Conditions : Haycrop just taken, ground dry.

Ref no.	Find	Details	Location	Depth	Comments
1	Buckle-breeches	c.1730–1780, perfect. Loop design 1"x 3¼". Pin intact.	Lane 1 3 paces	7"	Now in local museum.
2	Lead token	1500–1700 1" diameter Poor condition.	Lane 1 6 paces near phonebox	3"	
3	Button	1790–1810 Crown above lion. Standing on crown Military? Dragoons?	Lane 1 11 paces	5"	

Search ceased. Lane 1 complete, Lane 2 hardly started. Thunderstorm.

Alternative Methods of Plotting

Fixed Datum point — PYLON

1	By variation from Mag. North & distance from Fixed Datum point.	200° 60 metres	
2		210° 60 metres	
1	By 6 Fig. Grid reference.	748 942	

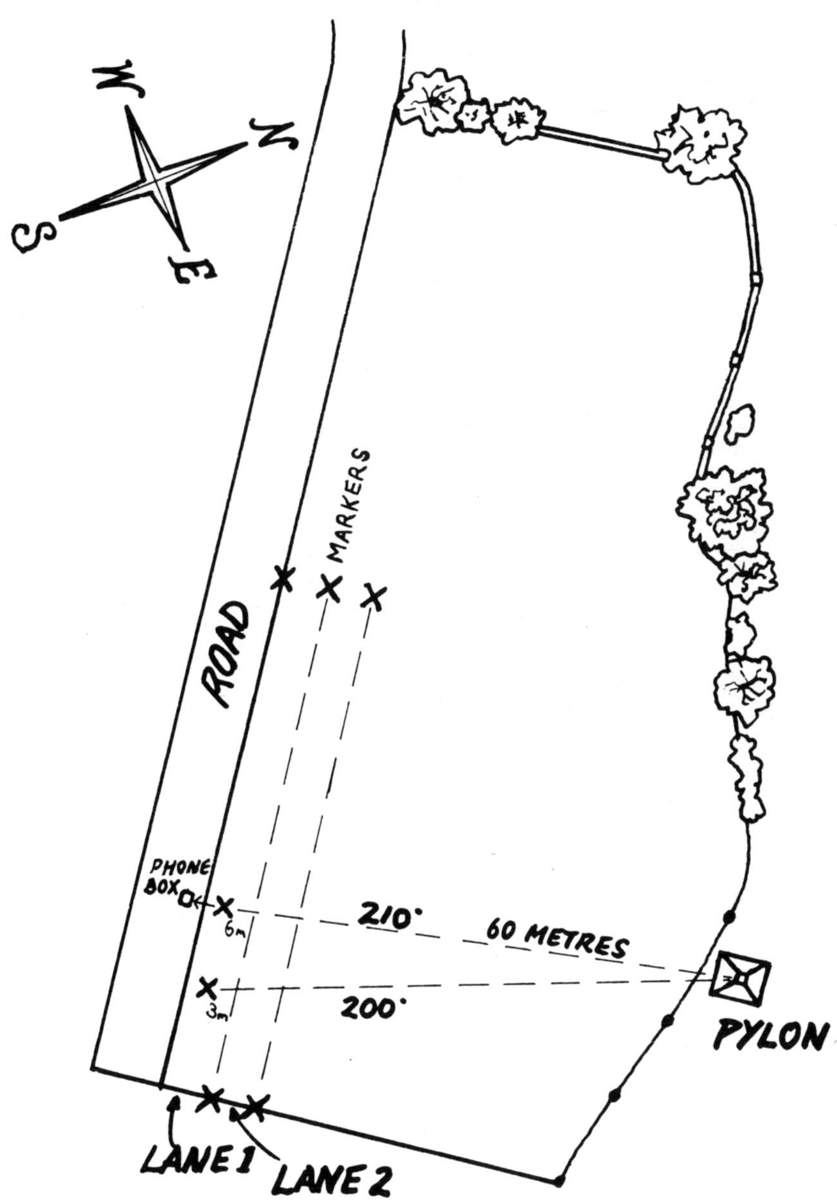

Map to show location of finds

Getting on with Archaeologists

From the point of view of the treasure hunter using a metal detector, archaeologists fall into two groups: those who recognise that metal detectors, if used correctly by responsible people, can make an invaluable contribution to archaeological studies in this country, and those who consider them the tools of the devil, set upon destroying our heritage by wanton acts of vandalism. Unfortunately, this second group seems to be getting a great deal of publicity and has been trying to get metal detectors banned in an increasing number of places. You are bound to come across archaeologists who feel like this at some stage so it is worth considering their arguments carefully.

First, what is an archaeologist? Basically, he or she is a person who from schooldays and through university has dedicated his life to piecing together the vast multi-dimensional jigsaw that comprises our past. Each puzzle piece has to be discovered somewhere under the soil of this country, cleaned, identified however tentatively, considered alongside other pieces thought to be similar and then interpreted. At any time a key piece may be discovered which will enable him to slot together a whole but still small section of the puzzle. He is fighting a losing battle and he knows it. At a rough estimate there are already enough Scheduled Sites in this country to keep archaeologists busy for another two hundred years, and these are only the sites which they know about; many other areas of great interest and significance must still be undiscovered. Time is running out, they are faced with a world in which the clues they hunt so passionately and steadily are being destroyed daily by the spread of cities, by the construction of motorways, by the use of deep ploughing techniques on farms and the loss of forest, common and moorland to increasingly intensive farming.

Then onto the battlefield strolls Joe Bloggs waving his metal detector in one hand and clutching a fistful of ancient artefacts from he can't quite remember where. Wouldn't you feel like stringing him up from the nearest lamp-post? I would.

It is a sad fact that it is the people using metal detectors irresponsibly who get all the publicity. The majority of metal detector users are ordinary, law-abiding folk who take great care to enjoy their hobby without treading on anyone else's toes. Increasingly, individuals and clubs are getting to know their local archaeologists, who are, after all, the source of a tremendous amount of information on the history of the area, its development, lifestyles, and the identification of objects found by detector users. But you are unlikely to read about joint activities in the national press – they are not considered newsworthy.

There are many areas where mutual co-operation could pay off. Have you ever seen an archaeological site which has had the top 'six' inches taken off by a bulldozer? A metal detector might have located any number of objects in there (most coin finds are within a few inches of the surface), which would otherwise be lost or destroyed by the heavy machinery.

As you begin to find interesting objects you will inevitably find yourself spending more and more time at your local museum. So get to know the curator. Take your finds along to him and he will help you identify them, especially if you have carefully logged the details of where you found them. It is a good idea to get a receipt for anything you leave with him (keeping a photograph of the object is even better), since he may receive a great many items and if he sends yours away for further identification it may be difficult for him to remember exactly which item was yours. If your find is suitable you may care to loan it to the museum to display. That 'Loaned by . . .' card can give you quite a kick – especially if like Tony Langwith's it appears at the British Museum.

And don't forget, there is absolutely no reason why you should not join your local Archaeological Society.

This archaeological site in Canterbury may not look very interesting at first glance, but a lot of painstaking and fascinating work is going on

Where *not* to Dig

Generally speaking, do not dig anywhere unless you are *positive* you have a legal right to do so.

If you are on private land it is not sufficient to have a tenant farmer's permission; you must have permission from the actual landowner.

The most obvious sites to avoid are registered archaeological sites. A new schedule of these has recently been brought out but copies are very expensive. Polite enquiries at the local museum, police station or Archaeological Society should be enough to ascertain these sites in your immediate area. But remember it is up to you to find out, for under English law, ignorance is no excuse. The most obvious spots to avoid are the sites of old castles, priories, Roman villas, etc. Churchyards are covered by ecclesiastical law, but not the area surrounding them.

Ministry of Defence property is, of course, strictly forbidden. An enormous amount of explosive material is still lying around in England on wartime ranges or washed up on beaches. Be careful wherever you are. If you find something suspicious *don't* hit it with a stick just to see what happens. One recent find included over 350 rounds of live .303 ammunition found within 3 inches of the surface and only 40 yards from a busy caravan site. Luckily the finder, a policeman on holiday, acted promptly, making sure no one else went near the spot and calling the local police, who in turn called the bomb disposal experts. So if you find so much as a suspicious white crystal powder in the ground, that is the procedure to follow.

Several councils are now forbidding the use of metal detectors. Your only course of action is to write to your MP and to your local council politely, but firmly, requesting permission to search. You must, however, be prepared not to be fobbed off with an inadequate excuse and a lot of patience may be necessary. It is best if you can persuade your local club to register their objections as well on the grounds that the louder the voice the more likely it is to be heard.

My next 'don't' is guaranteed to make some people laugh but it is still worth mentioning. Beware of the supernatural! You are probably not the type to worry about ghosties and ghoulies and things that go bump in the night and I, personally, have never had a psychic experience of any kind. However, this is a hobby which will tend to take you into out-of-the-way places on your own, perhaps for the first time in your life, and imaginations can work overtime under such circumstances. A number of people have reported being led to treasure by a ghost of some sort and they have obviously come to no harm from the experience.

Whether we like it or not a number of places have, over many years, acquired extremely unpleasant reputations. Beachy Head, for instance, is supposed to be haunted by the ghost of a black monk who lures the unsuspecting over the cliff. I know one hardened beachcomber in a lonely spot at dusk who literally dropped over £400 worth of equipment in the sand and ran like a bat out of hell for the nearest pub because he felt 'a cold black evil feeling almost as though something were drawing out my soul'. Safely in the pub he was regaled by the locals with grisly tales of murders and suicides on that stretch of beach and no one would go back with him until the sun was well up the next morning to collect his gear. Even taking his tale with a very large pinch of salt I know he has never been back to finish that beach. Incidentally, that was the only psychic experience he had ever had and he has never had another since.

Whether he was exaggerating or not, he obviously had a very unpleasant time and since we are metal detector users not psychic investigators my advice is to leave such places well alone. I took up detecting as an enjoyable hobby and that's the way I intend it to stay.

Where to Search

In spite of the restrictions imposed by the Government and some local councils there remains an enormous variety of places that are well worth searching.

The beachcombing metal detector user is now a familiar sight as he systematically quarters the beach when the crowds go home to tea. He, like you, has probably spent his day lazing in the sun but, unlike you before you acquired a detector, has made a very careful note of the exact location of the ice cream stall, the most sheltered spots on the beach and where that noisy coach party sat so that he can head straight for the places where he is most likely to find coins lost from pockets, rings and watches. One man in Cornwall recently found over 200 rings in less than a year – and almost all were in wearable (or saleable) condition. It is quite amazing how many people forget that their fingers shrink when cold or wet and rings slip off completely unnoticed.

The keen beachcomber will be out not only during the day but also all night long if conditions warrant it. The man who turns up at dawn after a night storm or spring tide may well find that other, hardier types have already cleaned up and gone home to breakfast. But *do* check the tide tables – and remember that in some places you can be cut off by the sea well before high tide. Check with the local coastguard.

Don't feel, though, that only a beach will do! Any site which is used regularly, or has been used regularly in the past, will repay careful searching. Footpaths, riverside walks, local beauty spots and lovers' lanes all contain their quota of finds – and don't forget the foot of slippery slopes: a man slipping may not notice coins, etc., falling from his pocket or may be too embarrassed to search for them!

Farm outbuildings can be highly profitable too, as Dave Kitchen found when he discovered his four golden guineas. Fields where old horse trading fairs were held are a rich source of treasure, so too are modern fairgrounds and circus sites – although you may have noticed that many professional fairground people are out with their own detectors before they leave. However, they are usually in a hurry, particularly if they search at night, and search very superficially, so they usually miss a lot. Take particular note of where most small change passes hands. Coins, even 50p coins may be lost or ignored if the loser is struggling with three ice creams, two candyflosses and a hot dog!

'Pick your own' sites, whether strawberries, apples or potatoes, could provide you with pickings of quite another sort. Hop fields can be virtual

Beaches are good places to search – so too are derelict buildings, often the source of coin and artefact finds

gold mines since the pickers were traditionally East Londoners on a working holiday. During the day they worked hard and money was continually changing hands. At night they played just as hard with enormous bonfires outside the rows of living huts and plenty of ale to help the party go with a swing. When the season ended they left behind a considerable amount of junk, but they also left coins, lighters, rings and a great many children's toys such as lead soldiers and boats, many of which are now collector's items.

Is your favourite holiday spot surrounded by trees? Good. Trees may sound odd things to search but they were favourite hiding places in days gone by. After all, many trees survive for hundreds of years and would have been easily recognisable when the owner wanted to retrieve his hoard. But life was often cut short unexpectedly, whether it was the highwayman getting his just deserts on the gallows, or illness, or violence, and many hoards were never reclaimed. Hidden, say, in the mighty but frequently hollow oak tree, these precious hoards may well be several feet above the ground, so search the bole of the tree as well as among the roots (being careful not to damage these, of course). I have an eye on an oak tree near my home which was part of the boundary between two parishes and alongside the route of a path from one village to another. Who knows, this could be *my* find of a lifetime.

Duck ponds and horse troughs, whether on a farm or in that pretty village where you like to stop for some light refreshments, have often been the scene of some fairly hectic horseplay among young men, with one or more of them likely to end up with a thorough soaking, especially at the end of a long hot day. During the Middle Ages the whole village would turn up at the pond to watch some poor unfortunate old woman 'tried' for witchcraft (if she drowned she was innocent, if she floated she was burned as a witch) or the village gossip ritually dipped in the pond on a ducking stool. With such entertainments to watch, and 'fairings' – pretty trinkets – to buy from the travelling tinker to give to your girlfriend, many valuable possessions were lost and quickly trampled into the ground. Many are there yet.

While we are on the subject of witches, did you know that they were often hanged at crossroads? So were highwaymen and other law breakers. In pre-scientific days hanging was often a protracted business and the bodies were often tarred and left on the gallows as a warning to others, attracting crowds of sightseers. Human nature hasn't changed. Modern catastrophes attract people in just the same way. How many times have you read in the newspapers that ambulances and fire engines have been unable to reach a location because the roads are clogged with onlookers? Such sites are worth searching – but only when decency permits.

Hollow trees and stumps can be fruitful places to search

Other modern sites which can be rewarding are parks, picnic sites and showground sites. Permission to search can usually be obtained in response to a polite request, a demonstration of your untraceable digging technique, and a promise to clear up all tin cans, bits of silver paper, bottle tops, etc., as you go along. (Again I must stress that *written* permission may save a lot of headaches later on.) Even more modern sites include new roadworks and demolition sites, civic dumps and those used by sludge gulpers. (You don't know what a sludge gulper is? Shame on you!)

Incidentally, thinking of an old site which I have been investigating recently, my local common, reminds me that recently two 13-year-old boys in Bedford searched their local common with one £40 detector and went home clutching a polythene-wrapped hoard of silver, including rings, brooches, a Victorian evening bag and some other items. The police believed them to be the proceeds of a housebreaking and appealed for information on Police Five on television. But no-one came forward to identify or claim the loot and in due course the ownership of the find reverted to the boys and the whole lot were auctioned on their behalf at Sotheby's sale rooms in London.

The Bedford schoolboys whose dream came true – they found a stolen hoard of silver on their local common

How to Research Sites

You don't like the word 'research'? Don't worry. You can do as much or as little work in this department as you feel inclined to do. A great deal can be achieved just by using your own common sense. Just thinking about your own behaviour and that of your friends can tell you a great deal about past times. For instance, trippers on a charabanc outing to, say, Southend, fifty years ago still had to answer the same number of calls of nature on the way home as a modern coach outing and they looked for the same sheltered spots and lost the same number of coins, cigarette lighters, etc.; and people have visited the same beauty spots, like Box Hill in Surrey, for hundreds of years. Whether they went to see a prize fight or a motorbike scramble they were equally careless with their belongings.

In the spring of 1978, Yorkshireman John Ball stood with his metal detector overlooking the whole Denaby Main valley. It's a big place and he could have spent years searching it, but he didn't. He used his imagination. He switched off John Ball and the twentieth century and imagined himself an old-time merchant worried because he was carrying too much money for comfort – not for him the welcome presence of a bank's night safety box or a well-insured hotelier's safe. He was heading for the ferry across the stream at the bottom of the valley and the warmth of the local inn on the other side. The problem? It was getting dark and ferries were favourite spots for ambushes by the all-too-familiar footpads. His solution? The money must be securely hidden and in a spot where he could find it later. Much of the valley is now defaced by mine workings, but John could see what looked like a natural outcrop of rock just half-way between him and the old site of the ferry. Convinced he was on to something, he went down and began to search. The result? Over 1,600 Roman coins. No book told him where to search.

A tremendous amount can be learned just by talking to the old man sunning himself outside the village pub. He'll remember where the old fairs used to be held, where the boys used to swim in the river or where the Peace Day celebrations at the end of the First World War were held. By listening to him you can learn a lot about the day-to-day activities of country people. For instance, did you know that an old cocoa tin with holes punched in it, containing a smouldering rag, would keep your hands warm for hours on even the coldest day while you were out on your daily chore of bird-scaring or gathering wood for the fires? Well, next time you find one you will know what it is; and there are even enthusiasts who collect old tins.

The only really essential information source is a copy of the local tide tables if you are beachcombing. These are available from piers and fishing shops. Make sure you read them and remember to adjust for Summer Time. Wear a watch, wind it up and remember to look at it frequently. It is surprising how quickly time passes and Davy Jones has enough treasure already without you adding to it!

Maps are obviously one of the most valuable sources of information. Old maps will tell you things such as where old villages were located and the footpaths between them. If a path crosses over, or even very near, a river you may be quite sure the local carter must have become bogged down or shed his load there at some time.

There are a number of extremely useful maps of different kinds. For instance, tithes (the payment of one-tenth of income) used to be payable to the Church in the form of all produce, whether pigs, eggs or cabbages, until the great Tithe Act of 1836. The Act laid down that payment should be made in money and every parish in the country had to prepare large-scale maps including every relevant detail of fields and buildings and exactly who owned them so that the right person could be charged the right amount. Most 'tithe maps' were prepared by professional surveyors who used scales of up to 80 inches to the mile; the smallest were at 6.7 inches to the mile. Three copies were prepared. One copy of the map and the accompanying descriptive list (the Apportionments) went to the County Record Office and the second was left with the parish church; the third copy was sent to the Public Record Office in London. These are now available at the Kew branch of the Public Record Office, though a special reader's ticket is required for which you will need identification. They suggest a letter from your headmaster on *headed* notepaper.

Where tithe maps are still in existence you have a statutory right to see them. Unfortunately, they are some 8 ft square and are not available in a conveniently reproduced form. You will probably need to trace off – in pencil, of course – the area in which you are interested so if you have a choice go to the source with the largest tables!

Ordnance Survey maps themselves date back to 1801 in many cases. Reprints of the 'Old Series' are available from David & Charles, South Devon House, Newton Abbot, Devon. The Ordnance Survey themselves have published certain old maps such as the Bodleian Map of Great Britain (AD 1360), Symonson's Map of Kent (AD 1596) and OS Southampton Area 1810. (A catalogue is available from them on request.)

Looking at old maps alongside their modern counterparts can be both fascinating and informative, and most local libraries will have some maps that you will find of interest.

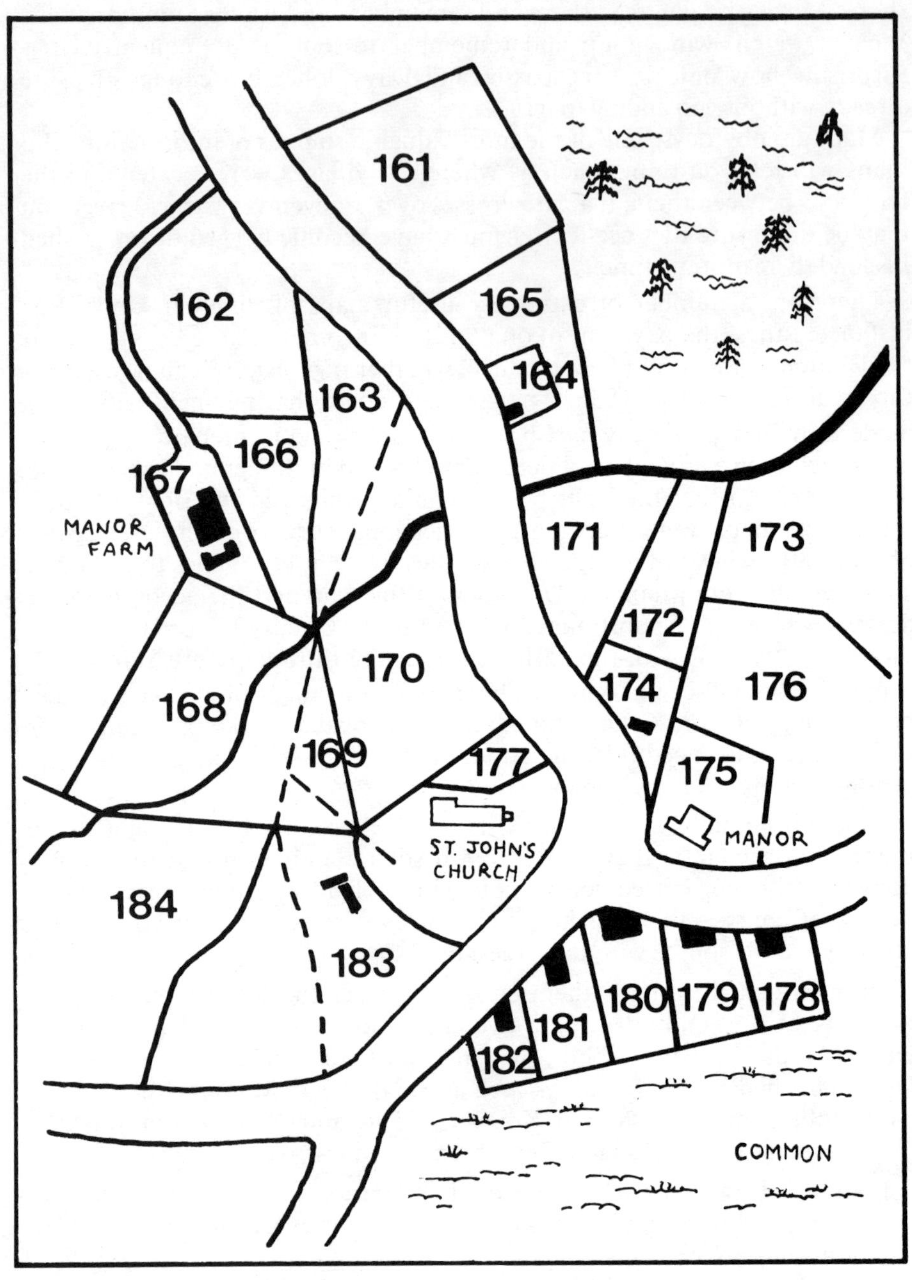

161
162
165
164
163
166
167
MANOR
FARM
171
173
172
174
176
170
168
169
177
175
MANOR
ST. JOHN'S
CHURCH
184
183
180
179
178
181
182
COMMON

Numbers referring to the plan	Name and Description of Lands and Premises	State of Cultivation
161	Lambs Field	Meadow
162	Hill Field	Pasture
163	Stone Meadow	Meadow
164	Cottage and garden	....
165	Plantation	Plantation
166	Big Orchard	Wood
167	Homestead, stables etc	....
168	Rubbish etc	....
169	Three-gate meadow	Meadow
170	Four-acre field	Arable
171	Clover meadow	Meadow
172	Narrow field	Arable
173	Rocky acre	Pasture
174	Croft and garden	....
175	House and garden	....
176	Waste	Waste
177	Church Field	Waste
178	Smithy and garden	....
179	Public House and garden	....
180	House and garden	....
181	Cottage and garden	....
182	Cottage and garden	....
183	Stables and pasture	Pasture
184	Legion's Meadow	Meadow

Opposite, a Tithe Map; above, part of the Award or Apportionment

Other sources of information, which vary in usefulness from area to area, include county annals, old diaries, parish registers, old guide books, historical society notes and local archives. The local library and Chief Librarian, and the Chief Archivists of libraries and museums, are all unparallelled as sources of information, and a wet afternoon can disappear in an eyeblink in old bookshops or in the records sections of local newspapers if you are lucky enough to be granted permission to search them.

Access to a list of 'scheduled sites' is necessary if you are to avoid those places which you ought to avoid. The List of Ancient Monuments in England is available through Her Majesty's Stationery Office. However, I suggest you first enquire whether your local library already has a copy and, if they have not, why not try to persuade them that they really ought to have one? You can then note down or photocopy the parts referring to your area.

Once started, many winter evenings can be spent reading your way through the innumerable books which may give you some ideas on where to search. Starting with William Cobbett's *Rural Rides* (published in 1830) and a host of books on highwaymen, country life and death, and continuing through heavier works such as *British Coin Hoards AD 600-1500* by J. D. A. Thompson (Royal Numismatic Society Special Publications, No. 1, 1956), the list is endless and each useful book will contain a bibliography of other useful books.

Laws Affecting You and your Find

It will be in your own interests to keep an eye on the national newspapers to see anything which comes up in connection with metal detectors and their usage. English law is a law of precedent: that is to say, every case is reviewed against the background of every other connected case, and since the hobby is still so very new each case just may materially affect your actions. However, most potential problems can be avoided with a little foreknowledge.

Since January 1981, Pipe Finder's Licences (which used to be a legal requirement for users of metal detectors) have no longer been necessary. The Home Office discontinued this condition specifically to 'lead to less bureaucratic control and to greater freedom to individuals'. Reading *Treasure Hunting Magazine* is a good way of keeping up-to-date with the metal detection legal scene, including the activities of the Detector Information Group (comonly known as DIC) which, spearheaded by manufacturer and dealer members, has helped oppose attempts by local councils to introduce anti-detection bye-laws. This group is recognised by the Home Office and the Department of the Environment and also runs its own legal advice service.

Next, make sure you have permission to search from the landowner, not a tenant farmer, preferably in writing. If you have permission to search Farmer A's land and have to cross Farmer B's land to get there, put your machine firmly away in its carry bag and do not be tempted to use it until you get to Farmer A's land. (The subject of permission to search is dealt with in more detail in the next section.)

If you are told you are breaking a bye-law, request politely to see a copy which legally must be displayed prominently. Have a look by the entrance to your nearest park to see what bye-laws look like.

The principal law affecting any finds you make is that of Treasure Trove. We cannot do better than to read the official notes issued by the British Museum.

British Museum Notes on Treasure Trove

Objects of gold or silver (including coins, plate and bullion) which have been hidden in the soil or in buildings, and of which the original owner cannot be traced, are Treasure Trove, and by law the property of the Crown.* It is of great importance for historical and archaeological reasons that any such finds should not be concealed, but should be reported and handed over in their entirety to the proper authority: a finder who fails to do this may be guilty of a criminal offence.

Anyone therefore who finds such objects should report them to the

Coroner, either direct, or through the local police, or by writing to the Director, British Museum, London WC1, who will communicate with the Coroner. It is always to the finder's advantage to report his find at once; for, as the law is now administered, he receives either the find back, or its full market value, as a reward for doing so. If the Coroner decides that more than one person was concerned in the finding, then the reward may be divided, but it should be emphasised that the reward is made to the actual finder(s) and not to the owner or occupier of the land.

When a find has been declared Treasure Trove it is dealt with in one of several ways: if it is not required for any museum it is not retained, but returned to the finder to dispose of as he will; or, if he so desires, the British Museum will arrange to sell it for him at the best price obtainable. If, on the other hand, the find or any part of it is retained for a museum, the finder is given the full market value of what is retained, while anything not retained is dealt with in the manner just specified.

Coins and other ancient objects of copper, bronze or any other base metal are not Treasure Trove; and finds of this nature need not be reported to the Coroner (though there may well be a duty to report them to the Police or to the owner of the land or building where they are found). The British Museum or the appropriate local museum is always glad to hear of such finds, and, if they are reported, may in suitable cases purchase them direct, or advise on their disposal.

Any further information may be obtained from the Director, British Museum, London WC1.

*Unless (as in some cases) the 'Franchise of Treasure Trove' has been expressly granted to a subject in so far as finds of a particular locality are concerned, and in the County Palatine of Lancaster. In the latter and in certain Duchy liberties outside the County Palatine, the Franchise is vested in the King in right of his duchy and not in the Crown.

In Scotland, all old articles newly discovered, whether or not they contain gold or silver, belong to the Crown and are treated as Treasure Trove. The position is similar in the Isle of Man where a failure to report the finding of any old object, whether of gold, silver or some other material, can be seen as breaking Manx laws and render the finder liable to prosecution. In Ireland, the laws are even more restrictive – in Northern Ireland, under the terms of the Historic Monuments Act, it is unlawful to dig for objects of archaeological or historic interest without a special licence. Any such find must be reported within fourteen days and if made of, or containing, gold or silver, will be subject to the normal British law of Treasure Trove. The regulations in the Republic of Ireland are similar to those of Northern Ireland.

Government legislation severely restricting the use of metal detectors by amateur treasure hunters on ancient monuments and historical sites came into force in January 1980. Fines of up to £200 await anyone ignoring this ban. Anyone damaging or destroying an ancient monument could receive unlimited fines and up to two years' imprisonment. Permission to use detectors on one of these sites has to be obtained from the Department of the Environment.

Fortunately, there is a tremendous amount of privately-owned land everywhere which we can search with the owner's permission.

Georgian silverware, the proceeds of a robbery, found near Catford

Landowner Agreement Form and Insurance Cover

Signed Landowner Agreement Forms are as essential as your metal detector. Getting one filled in and signed involves a bit of effort but is more than worth while in terms of avoiding later difficulties, such as the division of any money received for items found, not to mention avoiding any future legal unpleasantries.

There is no infallible method of getting permission. I find my best approach is to write to the Authority or landowner (remember a tenant farmer's permission is not legally sufficient), getting his correct name from the farming section of the local Yellow Pages, stressing that I wish to reach a formal agreement and am well aware of the need to protect his land, live-stock and property. I also give him brief details of my membership of the British Detector Society (any club membership also helps to lend an air of respectability) and my insurance policy through them which includes third party indemnity. If I have been researching the area I say so and try to drop in a snippet of information relating to that particular area. If the site involved is especially important to me I also enclose copies of the Landowner Agreement Form and Insurance Policy. Enclosing a self-addressed, stamped envelope is also a good investment. Naturally, I offer to split my proceeds with him.

However, I usually assume that there will be no response to my letter and telephone two or three days later. I refer to my letter and am prepared to use in discussion all the points put forward in this book: I have a rough idea of all the planting, ploughing and harvesting times in the area and what the weather has done to them, am aware that New Holland tractors are experimenting with pulse induction detectors on their machines, and am quite shameless in dropping the names of any local people of good standing on whose land or with whom I have metal detected (reputations spread as fast in this field as in any other), and am prepared to accept permission for a single day's searching with the farmer or his representative coming down at any time to make sure I am not devastating the landscape. In addition, I offer to remove from his field any metal objects I find which could damage his farm machinery.

More often than not my method works. You may need to persevere, however. On one occasion I was steadfastly refused permission only to be invited, nay begged, to come along when the landowner's wife lost her wedding ring taking hay out to the horses one winter!

Insurance is a subject to which you have probably never needed to give any thought, but many landowners are happier to admit you to their land if you can reassure them that you are covered. For instance, you accidentally leave a gate open and the farmer's cows get out into the road, causing damage to other land and perhaps even an accident. Or someone who smokes forgets to put out his cigarette properly and sets a field on fire. The cost to the farmer can be extremely high and certainly you and your guilty conscience will never be allowed to set foot on his land again – or any other land for miles around, for that matter.

I am insured with the British Detector Society's policy backed by the Commerical Union – one of the country's largest insurance companies. Not only am I covered by a broad form of legal liability cover for injury, illness or disease to third parties (that is, other people) or loss of or damage to third-party property arising out of the use of my metal detector, but also for accidental loss or damage to my detector and associated equipment. Should the awful day arrive when I am the one to be nudged into the pond by that herd of curious cows and foolishly get my metal detector control box wet, the detector would not be covered by my manufacturer's guarantee but it would be covered by my BDS policy. What's more, I would be able to claim, at today's market price, up to £300, rather than the price I originally paid for the machine. Should I be unfortunate enough to suffer spinal injury due to a fall whilst detecting, or lose an eye on a springing thorn branch, or even manage to kill myself off completely, I would also be covered.

The BDS policy is the only one I know of which is available to detector users from the age of 13 upwards, although on a restricted basis. For instance, death benefit for persons under the age of 16 is limited to £500.

This is a very important subject and I strongly recommend that before you start detecting you discuss it with your parents and your family's favourite insurance company, compare what they suggest with the BDS policy (details from the Chairman, 95 St Anthony's Drive, Chelmsford, Essex), and see for yourself where you can get the best value.

Competitors and spectators at the first International Treasure Hunt, sponsored by C-Scope Metal Detectors Ltd., and held at Stanford Hall, Leicestershire,in 1977. There was £1,000 in prizes, including a first prize of a metal detector. The competitors are searching for pre-buried tokens, some marked with a cash face value

Clubs are Fun

'Oh no, he's off again!'

How often have you said that as someone you know launches himself on a half-hour lecture on his favourite subject – or are your friends beginning to say that about you and your metal detector?

Nothing on earth will stop you talking about treasure hunting once you have got the bug, so why not spare (and keep) your friends by joining a club? Here you can talk (and occasionally listen) about the virtues of your machine, your methods, sites in your area and your discoveries. Of course, I can't guarantee that anyone will be listening to you, they are usually too busy talking themselves. Still, somehow it does not seem to matter. When you do pause for breath you will be able to pick up no end of useful tips from those who have been there longer than you. Many clubs fill their winter evening meetings with films, talks by manufacturer's representatives and lectures from local archaeologists and historians, and these are often both interesting and informative.

Clubs have mushroomed over the last year or so and many have junior sections, although the age limits for these vary considerably. A full list can be obtained from *Treasure Hunting Magazine.*

Any organisation seems to acquire a degree of respectability (or plain muscle power) not accorded to the individual. It follows, therefore, that many clubs are able to gain access to areas which are denied to the individual. You may well find, for instance, that your club has the sole permission to search local showgrounds after county or town shows, is allowed access to established council dumps or is invited to scour new sites marked for development purposes (you can probably locate old pipes of various sorts in return for the privilege). Once your club has built up a good reputation with local landowners you may well be invited to search parts of their land, say, certain interesting fields between harvesting and reseeding.

Rather more high-minded aims have motivated a growing number of outings by various clubs around the country. There has been a steady increase in the assistance they have been able to give the police in their investigations. My own local club recently turned out to help look for a latch key which was a vital clue in finding the murderer of a woman who had been found dead in her cottage in the woods.

The same club organised a sand-pit competition and display at an Essex County Show. Not only did they raise a sizeable sum for the kidney machine charity, but they also did very well by one schoolboy who found a pre-decimal penny in the sandpit. The organisers reckoned it was worth all of £20. Not bad for 5p!

Other groups with conservation in mind have been out clearing up river banks to make them safer for our wild life.

I greatly enjoyed attending a typical event organised by the Surrey Search Society last summer. Two large fields, tucked away in a lovely part of the county, were set aside for the competition. About a week beforehand the organisers went out to the site and salted away a large number of tokens fairly evenly over the whole area. It was a very hot day and quite a crowd of onlookers came with the competitors, determined to perfect their suntans while their fathers, husbands, sisters, etc., laboured under the hot sun. During the lunch break a representative from one of the detector manufacturers gave a demonstration of some of his machines (incidentally his two-year-old son was later seen hammering a tent peg with £300 worth of metal detector) and a mini-competition developed when club members tried out their own machines against his. The main event was late restarting as a result.

Competitions such as these are often split into a number of classes to give everybody a chance of winning something. Juniors and seniors, machines under/over a certain value, beginners, and even separate classes for men and women may occasionally be found – certainly the ground at the event mentioned above was successfully imitating cast iron and was enough to deter all but the toughest of ladies. Competitions run by manufacturers generally centre on classes for their own machines and they may even make these available to competitors – which is particularly useful if you have not yet managed to buy your own machine.

One side competition which caused great hilarity at the Surrey event had a number of different items – a horseshoe, a penny, etc. – buried not in the ground but under a thick rug. The object of the exercise was not simply to locate the objects, but to identify them correctly without seeing them. You can imagine the laughter at the expense of the know-it-alls when the rug is removed and they are proved to know less than they thought.

If there is no club in your immediate area it would be well worth while getting together with some like-minded friends and starting one. It can enormously increase the range of your activities and knowledge and make sure that you get even more fun out of your hobby.

TV star Faith Brown helps competitors at the first International Treasure Hunt search for pre-buried tokens – and finds one, though she doesn't seem too impressed with it!

What You are Likely to Find

If you ask a typical treasure hunter just what he or she has found you will almost always be shown a small display case of 'better' finds and a very large box of junk. As you will discover, anything you unearth yourself automatically becomes far too valuable to throw away. Every treasure hunter lives in expectation of the BIG FIND.

The single items you are most likely to find are *coins*. There is no short cut to recognising these. The only way is to look at them over and over again in museums or in private collections and to handle them whenever possible so that when you pick up a roughly circular piece of metal in a field your mental 'coin' bell rings. When you have washed it very gently with soap and water you will, as time goes by, be able to identify the most common types yourself, but please don't subject your find to the harsh treatment of a coin cleaner until you have got some reliable identification either from a genuinely knowledgeable friend or from your local museum, and have decided whether you are going to keep it for your own collection, in which case you may want it to be bright and shiny, or whether you want to sell it, in which case the patina of age is very important to a coin collector.

On closer inspection your 'coins' may turn out to be *'tokens'*. These were in fact issued to make good a chronic shortage of coins of the lowest denominations. They can turn up in an interesting variety of shapes. They are most likely to be dated 1650-1672 or 1780-1797. They were issued by tradesmen and usually carried the issuer's Christian and surname, his trade or occupation and the town or village in which he lived. Many bore the arms of his guild, or his family arms. Most of the earlier period tokens were for a farthing or a halfpenny, with pennies being quite rare. Silver tokens for a shilling or more came into use during the second period but they were extremely rare.

Victorian and Georgian coins

Early nineteenth century tokens:
An 1813 Birmingham Workhouse threepence and an 1812 Walthamstow penny

If you spend most of your time beachcombing you are most likely to find *coins and jewellery* – particularly rings which have slipped off cold fingers. If you are particularly lucky, you may find things which have been brought in by the sea, perhaps after a heavy storm; some of these may have been buried aboard wrecked ships for hundreds of years – even, for instance, since the Spanish Armada.

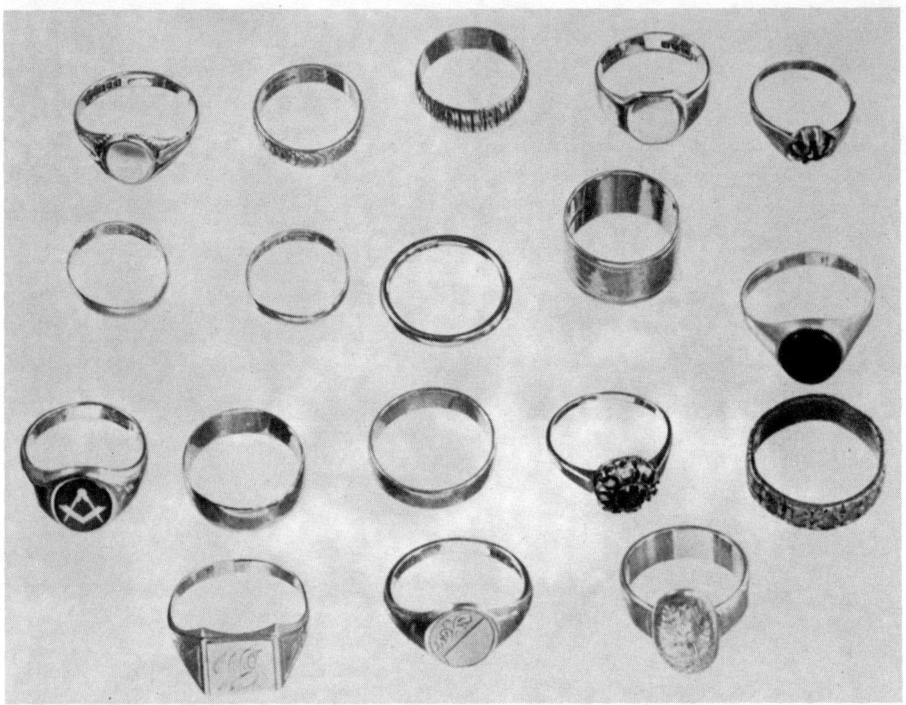

A collection of rings, mainly nineteenth and twentieth century

Some attractive jewellery finds

Searching in rural areas will undoubtedly bring you a fine haul of *horse gear* in the form of harness buckles, brasses, bells and shoes. Old paintings are probably the best source of information on these, although a substantial amount of information has been gathered together by enthusiasts such as Gordon Bailey, Chairman of the British Detector Society, and others for *Treasure Hunting* magazine. Generally speaking, buckles can be dated by their shape; Georgian buckles, for instance, were made in one piece and were rectangular in shape.

Georgian harness buckles and decorations

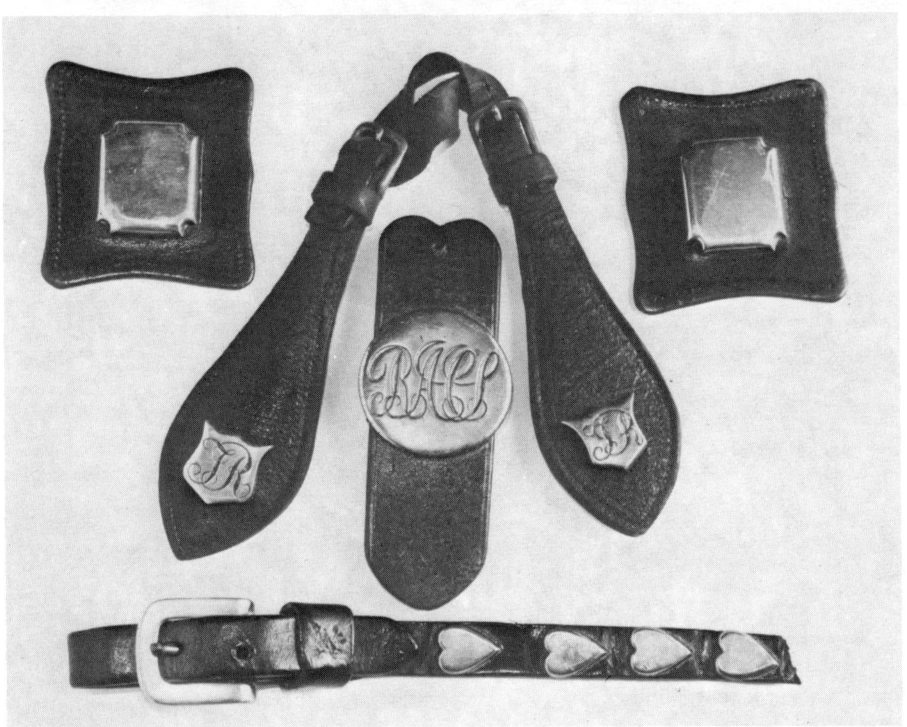

Horse brasses are too well known to need any detailed description. The oldest were very simple in design, sunflash or crescent being favourites. They were made by travelling tinkers and may still have hammer marks visible, particularly on the backs.

A collection of horse brasses

Horses have worn bells from biblical times onwards. The intention may have been merely to decorate the animals, or to ward off evil spirits or, more practically, to warn other travellers of the animal's approach. This would have been vital particularly after dark, since lights were very rarely carried, or when going along a one-lane track with high hedges on either side; and if you can imagine trying to back a bulky vehicle such as a loaded wagon, you will see why audible warning of another vehicle's approach can be useful.

The bells are generally round, but open bells were sometimes used. The bells were principally plaited into the mane of the horse, hung along its bridle, strung around its neck or suspended above its pack on a special frame.

There is a famous tale told by Horace Jones of Bridgnorth of a wagoner setting off for town with each of his horses sporting its own set of bells. It was very early morning and he had been up most of the night polishing the harness and the brasses so that when he set off it was still dark and he was half asleep. The ringing of the bells and the jingling of the brasses hid the fact that there was no wagon trundling along behind – he had completely forgotten to hitch it up!

Crotal bells

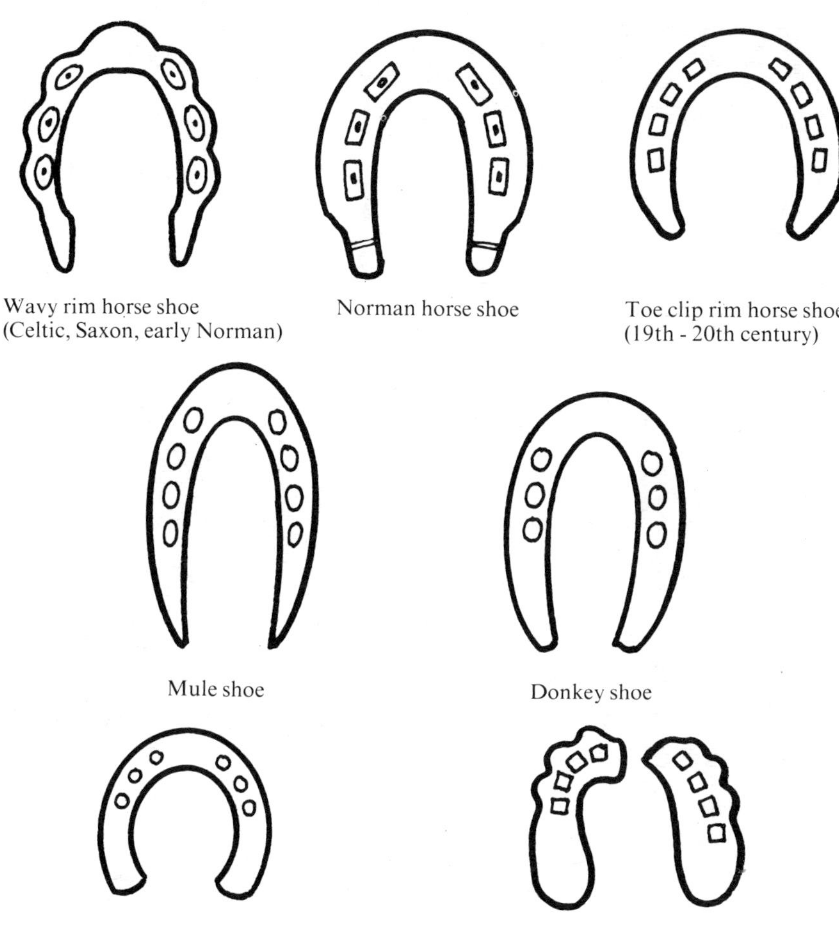

Shoes, of horse, donkey and ox, are something you may decide to collect out of sheer desperation since you will find so many. Ox shoes, of course, come in two pieces since they are cloven-hoofed animals. Donkey and mule shoes are longer and thinner than horseshoes.

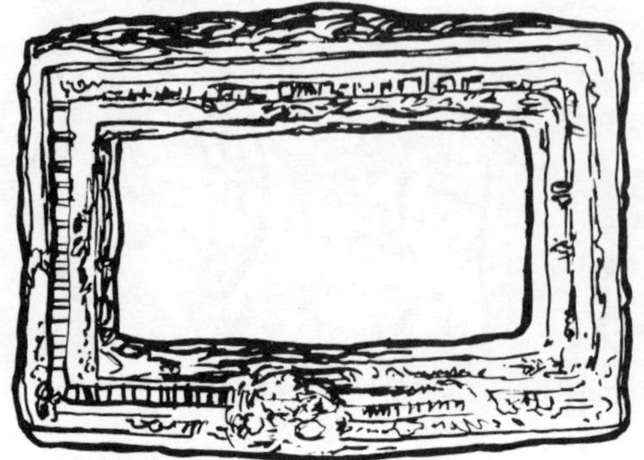

Large shoe buckle,
1770 - 1795

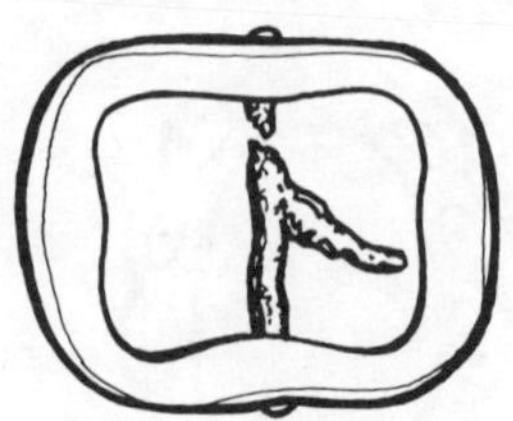

Shoe buckle, 18th century

Hat buckle, 18th century

Breech buckle, 1770 - 1800

Breech buckle, 1770 - 1800

Small spur buckle,
1660 - 1700

Belt buckle, 14th century

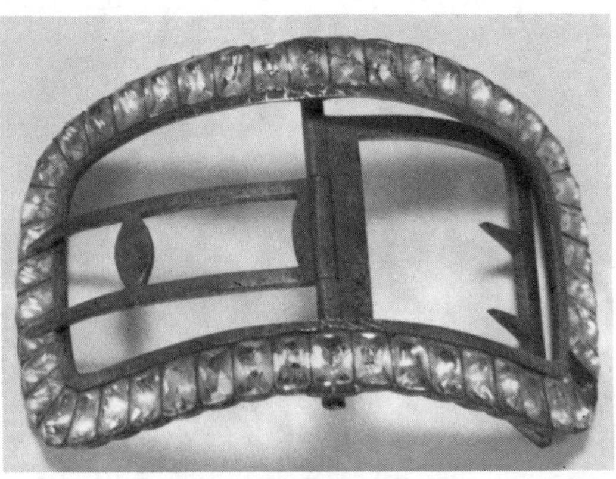

A Georgian shoe buckle

Shoe buckles (singly of course, never in pairs!) are something else you may frequently find. These vary from very plain tin buckles to highly elaborate silver buckles, according to the social status of the wearer. Other buckles may come from the bands of hats.

Buttons litter the countryside. Some are very simple round tin ones, but even these may have a history. Nearly every countryman had a 'hare pocket' or poacher's pocket made on the inside of his sleeved waistcoat on the left. Frank Whynes, a tailor at the beginning of this century, said: 'If you were wearing one of these weskits you could double up an owd hare and place it inside and button up: and no one knew anything about it!' Human nature hasn't changed over the centuries and no doubt fifteenth-century man had the same sort of pocket – which could be where your humble little button came from.

Horsemen – by which I mean ploughmen – were the aristocrats of farm labourers and very particular about their dress. Their waistcoats were frequently fastened with horseshoe-shaped buttons, cheap ones for everyday and better ones for their Sunday walking-out dress. The same shaped buttons were also used to trim the outside of their Sunday trousers, which were shaped very much like a sailor's bell bottoms. Buttons and buckles from the gentry are fairly easily recognised because of their obviously superior quality and because they are more recorded in portraits, paintings, etc.

Some types of button:

Military buttons may turn up anywhere, even in the strangest places. Prior to 1855 there was no uniformity of design but family or county crests and regimental numbers seem to have been favoured. After 1855 a regulation brass button, carrying the arms of the House of Hanover, was introduced.

Royal Welsh Fusiliers   'Staffordshire Knots'   Royal Irish Rifles

Argyll & Sutherland Highlanders   Royal Scots Greys   9th Lancers

17th Lancers   Royal Artillery   'Wolfe's Own'

Every army regiment also has its *badge*. This is worn on the collar of an officer and the cap of the ranks. Cap badges are obviously larger and more numerous and, therefore, more likely to be recovered with a detector. They can usually be recognised by having a vertical slide fastener or a double loop fastener secured by a split pin.

Military badges

Increasing interest has been shown by collectors in old *weapons*; swords, knives, muskets and old hand-knapped arrows and, more disturbingly, caches of old ammunition still in a highly active condition, are regularly being dug up. As long as you take no chances these can make up a fascinating collection. Don't forget, however, if you wish to keep modern firearms, usefully defined as any firearm which breaks at the breech to take a cartridge, then you must obtain a firearms certficate.

Other modern weapons in the form of highly expensive aluminium arrows are always getting themselves lost and many clubs will be only too happy to avail themselves of your offer of searching for them after an archery event.

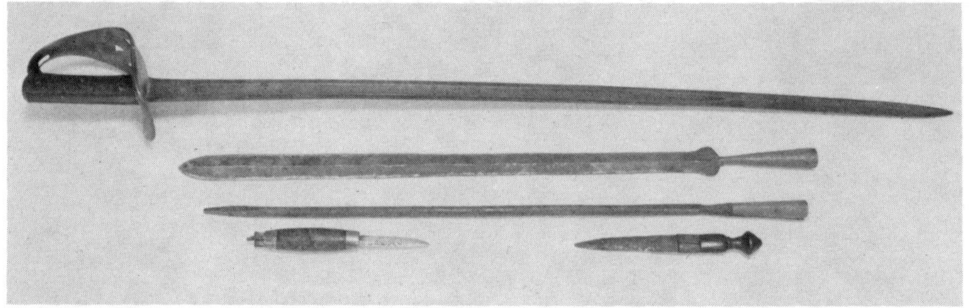

Some old weapons – knives, spears and a late eighteenth century sword. The knife at bottom left, from the late nineteenth century, is one of the first penknives

Old *farm implements* nowadays fetch premium prices at the many farm museums which are growing up throughout the country and may well repay the considerable effort involved in digging them up – and tidying up the site afterwards.

Aircraft may seem a pretty unlikely thing to be searching for with a metal detector, but only recently a Wellington bomber which crashed during the last war was discovered in East Anglia, and other aircraft have been found complete with the corpse of the pilot. These should be reported to the Aircraft Preservation Society who will pass the information on to the Ministry of Defence.

All these things are very interesting but we would be less than human if we did not all have in the backs of our minds the feeling that one day we are going to make the FIND OF A LIFETIME. I am not too fussy about what I find as long as it is legal and is worth a great deal of money. Four gold guineas worth £7,500 will do, but I really have my mind's eye on bigger game.

King John's treasure lost in the Wash is the most obvious choice, but I would be prepared to settle for something a little less famous. Old Mother Redcap's treasure was said to amount to £100,000, amassed whilst keeping an inn much frequented by sailors in New Brighton, Cheshire. It is said that she buried the lot in front of the inn bearing her name just before she died. The location of the inn is known but of the treasure . . .?

Dick Turpin is reputed to have buried several hoards in and around London and one is said to be buried near the Rose and Crown public house at Clay Hill in Enfield, the inn that was run by Turpin's grandfather.

Undoubtedly my favourite find would be the Druids' Hoard hidden in the famous Chislehurst Caves in Kent. The Druids are supposed to have gathered together a great hoard of gold and silver when their order began to be persecuted, and one of the last surviving members of the order claimed it was all buried in Chislehurst Caves in a passage 'where there are three small holes close by each other and thereunto a rock which hangs from the ceiling to the length of a man's arm' – not much help in caves some twenty miles in length.

My 'local' treasure is said to be at Great Baddow. That particular area is rapidly disappearing under housing estates but I'll keep looking!

Chislehurst Caves – a hoard of Druid treasure waiting for you to find?

Ready For the Off

At last the great moment has arrived and you are ready for the off. Last night you checked your equipment and maps, your batteries are all OK and your boots are not leaking. You've decided where to go and how to get there. You have your Landowner Agreement Form in your pocket together with details of your insurance.

If you are going on mountain or moorland you have checked the weather reports, left details of where you are headed, the route you intend to take and when you plan to be back. Your survival kit is packed and your tide tables for beachcombing are up to date.

You have read this book and picked up your lunch. Your family are ready and waiting – whether to come with you or to see the back of you! Your head, like mine, is stuffed with thoughts of pirate hoards, gold coins, bronze buckles, muskets and buried aircraft. Perhaps we won't find them today but tomorrow we are bound to make the FIND OF A LIFETIME.

Good hunting!

A Code of Conduct
As issued to members of the 'Treasure Hunters' Association

Amateur treasure hunting with metal detectors is growing into one of Britain's major outdoor family hobbies. And rightly so. It can be enjoyed by young and old; it combines fresh air, fun and an awareness of history with the exciting possibility of valuable finds; and it doesn't cost a fortune in equipment to get started. Which is wonderful – until we start treading on other people's toes.

1.   Don't interfere with archaeological sites or ancient monuments. Join your local archaeological society if you are interested in ancient history.

2.   Don't leave a mess. It is perfectly simple to extract a coin or other small object buried a few inches under the ground without digging a great hole. Use a sharpened trowel to cut a neat circle; extract the object; replace the soil and grass carefully and even you will have difficulty in finding the spot again.

3.   Help keep Britain tidy – and help yourself. Bottle tops, silver paper and tin cans are the last things you should throw away. You could well be digging them up again next year. So do yourself and the community a favour by taking all rusty junk you find to the nearest litter bin.

4.   Don't trespass. Ask permission before venturing on to any private land.

5.   Report all unusual historical finds to your local museum and get expert help if you accidentally discover a site of archaeological interest.

6.   Learn the Treasure Trove laws and report all finds of gold or silver objects to the police. You will be well rewarded if the objects you find are declared Treasure Trove.

7.   Respect the Country Code. Don't leave gates open when crossing fields and don't damage crops or frighten animals.

8.   Never miss an opportunity to show and display your detector to anyone who asks about it. Be friendly. You could pick up some clues to a good site.

9.   If you meet another detector user while out on a hunt introduce yourself. You could probably teach each other a lot.

10.   Finally, remember that when you are out with your detector you are an ambassador for the whole amateur treasure hunting fraternity. Don't give us a bad name.

Acknowledgements

The Authors and Publishers are grateful to the following for permission to use copyright material:

Keith Cogman of ICA Studios: pp.76, 79 (2)
Marguerite Fuke: pp.11, 21, 26, 50 (2), 65
Tony Garnett: p.89
Roger Johnson: pp.9, 11, 14, 17, 21, 26, 28, 33 (2), 46, 50, 51, 57, 73, 81, 82, 83, 84, 85, 86, 88, 91, 92
NPC Publications: pp.61, 63
Alan Powell: pp.68, 69
Jeff Short: pp.11, 22, 36, 61

The publishers would also like to thank Greg Payne and Roger Smith for their help in preparing this work for publication.